# A
# NEW
# COSMIC
# RACE

MAURENE WATSON

Order this book online at www.trafford.com
or email orders@trafford.com

Most Trafford titles are also available at major online book retailers.

Print information available on the last page.

ISBN: 978-1-6987-1151-5 (sc)
ISBN: 978-1-6987-1150-8 (hc)
ISBN: 978-1-6987-1152-2 (e)

Library of Congress Control Number: 2022905030

Digital Art: www.adeleandmichael.com & www.itstimetoawaken.com
Phone - 267 421 6667 / 215 680 2351

*Trafford rev. 03/16/2022*

www.trafford.com
North America & international
toll-free: 844-688-6899 (USA & Canada)
fax: 812 355 4082

# INTRODUCTION

## *Metaphysics of Love –*

*Welcome, Masters of Metaphysics and Bio-Essence Love, Light Energy Communication Masters, Energy Potential Magi, Love and Life fulfillment Masters, and those Light Beings awakening! You are now illuminating, transmitting, and streaming forth light fusion in a new standard of consciousness for a new cosmic race; through your free energy mastery of the biology and bio-essence of love and its metaphysics. Divine Self moves life's heart's bio-essence from the human to the soul's light body transition into the spirit's ascended form, and into the stellar/star-sun biosphere, which is a metaphysical or meta-sense form. This stellar star-sun biosphere is an illuminating Heart Sphere of light, which allows you to explore the new infinite unknowns of light fusion, to be lived as stellar beings of light; where you can light-travel, imprint-manifest, and create in your heart awareness instantly. This includes access to the higher realms, sub universes, windows, staircases of light, and super universes beyond the beyond into the infinite unknown of pure potentials. This light fusion as a divine human includes energy dynamics of your own unique soul-spirit consciousness. It is the shift of light body to an ascended bio/imprint or form-vessel, and beyond; into its growing stellar biosphere. Here in, you explore your own unique metaphysics and bio-essence meta-sense Love in infinite unknowns and as part of the new genetic/cosmic race. This is only/always within your own conscious awareness and energy or illuminating Heart sphere of light. Herein, light[fusion mastery over divine-human biology DNA/cell love, allowed the soul's bio-essence to experience and grow all life forms through self-love, self-acceptance, and self-awareness into a new standard or genetic ethic of love.*

*The gene of compassion has built in genetic integrity in the DNA-master soul code Essence. It has monitored the growth of the soul so it could be grown, rewired, re-spliced, re-essence/ed, restored, to master bio-essence cell love in free energy. This homeostatic dynamic returns any distorted essence back to its natural state; for its journey into the new light fusion realms, worlds, and universes. Therein, any alien: distorted, unnatural, or trapped energies could resplice, regenerate, and adapt all species life codes for the new cosmic species of light fusion.*

*Indeed, your universe's experiment, exploration, discovery, and next journey seeded as one new cosmic race, throughout the cosmos, came from mastery over the bio-essence of grown love. As these new galaxies, worlds, and light universes appear in your awareness, you will realize they are inside your own illuminating star biospheres as your own infinite unknown potentials to experience super-universal light fulfillment; which includes the ascended Divine heart-essence-human prototype you mastered. Your awareness is a base standard of soul heart essence mastery in self-acceptance, self-love, and self-realization, allowing Essence Heart to change atomic-quantum matter in any form of life via light fusion.*

*So, why is any soul body, form, cell, imprint created so <u>important?</u> This is so, because the forms, codes, blueprints of creation, when: separated, fissional, or fragmented by their creators; allowed creational forms to become a battleground for distorted power over the essence master codes to control existence and all the information within it. It created an illusion that the Presence of Life, or all Creation, could be controlled, forced, or hijacked; against its DNA-code; to prevent Soul's free energy willingness, (free will), to evolve. However, the new* *light fusion's vessel standard bio-essence of love with its fully grown metaphysics; or meta-sense essence attributes and gifts of creation, <u>restores genetic integrity.</u> It also prevents any soul-essence form, body, imprint, or bio-organism from ever again being occupied, enslaved, programmed or controlled by an alien or distorted, illusions, or foreign energy outside one's soul consciousness and energy! Heart Light's fusion illumination is its own protection. This allows the stellar biosphere to create any new life form, universe, imprint, and experience its fulfilment, joy, love and play; and then release it back into free energy when the heart essence is full. Thereby, it remains a new potential transmitted, shared, and imprinted as a gift to The All of life and the Cosmos!*

*Creation's infinite/unlimited imagination, inspiration, creativity, and joyful play via the heart reveals the difference between Old Earth-limited creation-communication energy and New Earth-unlimited creation-communication energy. Light body or ascendant inspiration comes from an essence heart standard of love. It's Isness can be quiet in creative stillness or regeneration. It can also be in its active energy play. Life cares for life. This Essence Heart, is just 'being' in creation, and inspiring and receiving its inner light fusion love illumination. It registers no heart posturing, mimicry, or imposing/harmful energies on anything or anyone. Spirit heart inspiration then, can be triggered by: imagination, an experience with your Divine Presence, communication with a child; by joy of a walk, a blue sky, the smell of a rose or perhaps a light travel experience communicating with the higher cosmic realms. Inspiration has a meta-sense awareness quality always arising naturally from the deepest consciousness of your essence-heart. Here no passport or justification of existence is required because the human soul and spirit wisdom experiences have been absorbed back into the light fusion embody to be used in new applications from your growing master code imprints.*

*Your wisdom and embody of lifeworks then, are about potentials in creative manifestation, creative abundance, and creative play. This invokes the 'Spirit Child' within the all of life with creative new careers within creative life fulfillment, and creative life relationships. You don't have to work at it. It is you, and comes to you, from your natural energy ethic of Essence heart's mastery over love in each DNA stem cell; as the maturing of bio-sense-essence love. Such energy awareness honors and respects your own feelings, senses, and essence in self and in others as they arise. This means allowing the human emotions, the angelic soul senses, and the childlike spirit essence that is the divine human to arise and guide. This is the natural Self, with no chasing the next or justifying; but connecting deep within the moments in the heart that guide your light fusion's illumination. Then life builds itself for you, and serves as an authentic living illuminating example, to all others in their life's awakenings and potentials. Here, there are no posturing or fake hearts; or controlling and judging in self/other, by not allowing or honoring the momentary old feelings of depression, aloneness, hurt, rejection, or denser feelings that mastered the beauty of having been a human; and taught the spirit how to love and grow the gene of compassion into the bio-essence organic of love.*

*Most Old Earth human body imagination, inspiration, and creativity came from human limitations of: mind-fear about safety existence, competition, control, feeding energies of external power agendas to compete; or be accepted by a mass programmed and limiting unconscious standards of success. The mind could rarely be quiet in the heart enough to feel that just being in creation's existence and accepting all life has to offer without judgment, was the true secret star to fulfillment. The justification of ones right to exist constantly trapped free energy of the soul. You are now well aware of the gravity entrapment drag of negative thoughts feelings, attitudes, and beliefs that needed the rebirth of awareness to integrate and re-embody all your human, soul, and spirit aspects that heart experienced; to feel self-love's fulfillment. Here in, you have light fused, rebirthed, resurrected, and grown the new heart essence species with a, regenerative divine-human/DNA code. Thereby, soul's core light vessel has deleted any fission or fractured light, that caused ancestral distorted alien or DNA that: would continue to impose an animal, violent, or of sub-human nature on the newly birthed humanity and all life's species arising in fusion crystal-diamond-star-sun light. And, any shadow, disgruntled, or imposter hearts that intend false power or harm are: exposed for their responsibility to heal, screened out by their own energies, and returned to the source that created them, or superseded by light fusion networks.*

*Therein, the standard of consciousness for discernment of love's light fusion and illumination has a built-in higher standard of how a master uses their own consciousness and energy. The Human-Master Heart quite naturally does not yield to old energies of psychology wound, mental analysis, an imposing or imposter heart, or interference on another soul's journey. They do not infringe, judge, heal, fix, or impose their energies in any way unless asked to illuminate potentials for another's light in another's awakening; therein triggering their soul heart's inner light fusion communication. However, they may collaborate in sharing new conscious transmissions or opportunities with other light networks and cosmic beings and universes who are aligned with and part of Humanity's light ascension. Overall, the use of their human-Master consciousness and energies are free to open up their own stellar star-sun corridors, windows, staircases, light universes, and infinite unknown of potential that are of greater service to all. This is so because each master-heart consciousness first experiences these visions, innovations, transmission assistance, and potentials; and meta-sense realizations within themselves and their own bio-essence of love first. Then and only then, are they authentic in their master soul's light illumination across worlds and the cosmos. Again, it's master* heart comes from their radiation of growing light. It magnetizes and amplifies the soul light of those ready to receive or be triggered by such ever expanding light, experiential love, and triggering of the innate bio-essence natural gifts. These new infinite unknown potentials in the master code imprints of Divine-soul's bio-embodied potential are then available to all humanity and the All That Is!

A Love-Life Heart Master's continued growth in their heart's light fusion potentials and illumination becomes their greatest fulfillment and gift to all humanity; replacing the old way of suffering service work, energy, holding, or empathic enabling of another's energy responsibility. This assists the New Earth-Star Gaia to fulfill its role as a cosmic spaceport for both the light body, the new ascended form, and the stellar/star-sun biosphere of sovereign light fusion illumination. It also assists each master to continue to fulfill its massive potentials on the New Earth and the new galaxies and universes that will be appearing in the super universes of light; which go beyond ascension into each soul's infinite unknown of potentials. This is because light fusion mastery over the human biology and bio-soul-essence of love; allows the heart essence to operate its _imprinted heart cell_ as a: transporter star gate, a magnetic imprinter, Source Code/r, centrifuge, quark stem cell particle and bio-ship for New Earth spirit matter, inside embodied love? Remember, life is a heart fulfillment center.

So, as, Gaia Earth moves fully into her light fusion Star-Sun vessel, your planet also serves as a cosmic_spaceport_for light body transport, the ascended form, and beyond into your own stellar biospheres with the heart as its own _mobile stargate_and center of gravity. Your interstellar space telescopes and craft will soon discover this, as well as; the supporting life from all your interstellar neighbors and all the new cosmic race universes awaiting your next journeys discovery and exploration into the cosmic age of light.

*In sum, we remind again and again, that in the super nova star-sun shifting of New Earth, Heart's biosphere's awareness returns as a: new experience, a new potential, a new manifestation, and a new experience of bio-essence love. And, when that moment/creation is experienced and fulfilled, it dissolves back into free energy! In your own biosphere a moment is as a world born, experienced, enjoyed, transmitted throughout the cosmos, and released back into essence. And, so it was when you first came forth from creation. Heart awareness frequency, is/and equals, instant manifestation of heart's joy, fulfillment potential and creativity as the natural bond to your own creation as the Divine being you already are simply by allowing your stellar biosphere to shine its light illumination wherever it heart awareness travels, creates, or manifests its soul imprints in the cosmos.*

*Light fused creation, not fission creation, inside your continued conscious heart's self-realization, will then explore how your new cosmic race and all your: crystal, plant, animal, and hybrid species live in the light? What clothes, homes, lifestyles, foods, light vessels, technologies, relationships, families, and new interstellar life systems will you design on your planet and with your stellar neighbors. You can avail your beautiful planet as a spaceport for such light vessel transport; where you can access travel in your heart awareness instantly. Notice, how atomic-quantum particle light fusion, is rapidly fusing space-time, and all converging past-future time lines/ existences to integrate and heal, and allow planetary and comic migration each soul light's fulfillment. Its language is transmitted in your new careers and lifestyles as in: designer imprint/replication technologies, plasma magnetics, Astro-sonics, interstellar commerce, intra-world or stellar inter-species sociocultural communication networks, interstellar contacts assistance and collaboration; as well as cyber-sonic spaceport systems, or lifestyle mentoring and fulfillment centers. Enjoy your own unique metaphysics and bio-essence meta-sense Love, where Heart awareness Is!*

# EXERTS

Energy Communication Masters, as living examples, you will now embark on a journey in the next six decades; to guide and illuminate humanity as they become their own free energy soul-masters. As such, the embodied heart essence must be free energy. **Energy Light-fusion Masters, the new heart consciousness in the Divine-human light vessel transcends all physics, science, and technology.** And, your Cosmic Heart experiences the awareness of that freedom, inside each embodied Master Soul Heart Essence-DNA code imprint. Free energy is Essence Heart guidance communication moment to moment potential to potential. Heart vibrates essence matter into existence and it simply appears into your hands and use. Your beautiful heart knows what would fulfill its every potential and **simply vibrates it into awareness**; because it already exists in The All That Is, or Isness of Creation. Your quantum light body instrument also serves as an <u>**adaptive imprint**</u> for your new species evolutionary organism that is evolving all life for all the ensouled children of creation. Indeed, the Sovereign Heart-biosphere will continue to adapt for all the cosmic races until a new **race of Peace** appears! Embodied soul Essence experience of the uniqueness of the Oneness; or genetic multiplicity within diversity, ascends enlightenment, back into a seeming mystery **of a meta-Essence Heart.**

Today we begin Part I of transmitting the essence energy imprints of the next generations of light children. The life code of their career soul-designs lives in the creativity of their heart's light vessel. Their soul-code imprints, in their core light essence; carry the qualities, tones, hues, and vibrations of the new light careers and lifestyles, they will live in the light vessel. Leadership in the light vessel, is living in the creativity of who these light children are, as a Heart Essence Being. They answer to their consciousness and their own evolving potentials, which manifest into expressive forms. However, they will be using the New Earth consciousness standards you Light-Energy Masters have anchored for them by being living examples in your light vessels. Your lives are the authentic stories of those who have walked before them. They don't want agenda leaders or lecturing rules, or dinosaur hierarchies; but those who understand, support, or choose to mentor them, in order to share their own unique-creative light gifts with your worlds. Most of them will **design their own careers,** yet unnamed, as they share their soul with life and humanity to fulfill their journey on Earth School and move into the Super-Universes of Light. The density of the animal spirit senses used here, help describe the merge of their: human emotions, angelic senses, and spirit essence blend; that integrates the new Essence Heart-DNA Master Light vessel. The light children have access to, all or mixtures of these meta senses described herein; which creates an adaptable model for the Divine-Human prototype for new paradigms in the New Earth light cycles. The **purpose** of mastery of the light vessel in the next generations **is** that it will end the need for the reincarnation in the coming light universes. This is because, Light-vessel's Essence Heart DNA codes can imprint any form it chooses, to experience through the essence blend qualities of **quantum-density**. These cycles could accelerate based on the overall consciousness of humanity, and critical mass ratios in the growing adaptability of the light vessel.

*New Earth remains a genetic universe and is being fully restored to genetic integrity. It's all part of disclosure and the truth of who you are as a species and what your IAM DNA carries in your bio-physicals. Your fully conscious bio-physicals, along with Gaia are seeding all the new Quantum multi-helixes. These include the new light bodies as well as cosmic intelligences or quantum codes to build new worlds and create with dark matter. You are the Meta Universal School that you have all become.*

*Masters, light body is your Divine-Human spirit embodied in quantum density. Light vessel goes beyond physics, technology, and science. It will evolve its DNA codes and transcriptions exponentially throughout the many New Earthlight super-universes. The Essence Heart is your: transporter star gate, a magnetic imprinter, Source Code/r, centrifuge, quark stem cell particle and bio-ship for New Earth spirit matter, inside embodied love? The light body in the Multi-light Universe is a blend of the physical and nonphysical into new conscious superconductive light systems. These bio-systems include new adaptive DNA Source bio-soul code templates made of organic essence consciousness.*

*Its heart cell is a blend of a: crystal soul cell, a diamond spirit cell, a multi-plasma orb, and liquid light particle cell. It is a new heart stem cell that can regenerate, re-imprint, or repair your entire bio-organism right out of your own consciousness. It is a blend of Old Earth atomic and New Earths quark blueprints. It is a blend of Linear and multiple applications of time and space.*

*Your heart Consciousness is a source code imprinter, tracker, super essence sensor and information scanner. And as a lover of life in all its aspects, relationships, and life experiences; all is available in a blended new sensate of unique chosen reality experience. This has always been inside natural receivership of life itself; as your own full conscious light is more deeply embodied than ever before. In this new matter, all the quantum super/meta multi senses have new qualities of Essence light you have grown in the DNA template blueprints for all life forms* and species. *What is it like to walk in a vessel fabric of plasma light matter? What does light particle fusion: taste like (soma; sound like, (clairaudience); look like, (clairvoyance); touch/sensate like, (clair-essence), travel like (tele-transport) communicate like, (tele-commune), inside your own consciousness; where nothing happens <u>unless your heart chooses it</u>!* Essence light, as your bio fabric and multi-senses, replace any old mental, emotional, physical, or spiritual addictions or obsessions of thoughts, feelings, or beliefs. *What is it like to imprint your consciousness on an object, an idea, a passion, a cell, or on a new experience? What happens when quantum particles disappear and reappear? What happens when matter can change its own essence through freeing itself that it might interact with life in any way it chooses?* This new inner contact allows for a constant dialogue and conversation with the cosmos in all the spheres of quantum light. This is not soul extension where the embodiment dies in an unconscious state or is locked in a non-physical existence. Rather this is soul infused essence embodiment in conscious fluid transcendent states of change. *What would pink compassion liquid light water sense like?*

# ARTICLES

*Q: How does light body create an abundant life in the*
*Multi-D Master and change Empathic perfection stress?*

# The Story of Love and Creation

*~ Walking Life as a Master in the Love Body ~*

## *Maurene Watson*

*Edited by Susan Mary Gardner*
*Artwork by Maia Christianne Nartoomid*

# In Your Genetic Universe Female DNA and Male RNA Emotions, Senses, and Essence Bio-merge into Divine Heart

New Earth remains **a genetic universe** and is being fully restored to genetic integrity. It's all part of disclosure and the truth of who you are as a species and what your IAM-DNA carries in your bio-physicals. Your fully conscious bio-physicals, along with the bio-soul of Earth are seeding all the new Quantum multi-helixes. These include the new Essence DNA vessels and cosmic intelligences or quantum master codes to build new super conductive light systems as worlds created with dark matter. You are the Meta Universal School that you have all become. This is because full conscious embodiment is returning full Essence genetic integrity to all soul contracts again. The timelines of ETs or inter-D Oligarchies using humans as their genetic Petri dish is over. When you opened the new Quantum fields, you removed the artificial intelligence or **_AI Chip_**, from the Solar Plexus of the Universe. This **_synthetic mind_** had grown a tumor feeding off essence in every human life system and activity. That is why it has taken so long for us to restore, code, and grow the Essence-Light bodies into the Divine human prototype. The Ancients and Elder Councils did not need ET cloning. Creation progenitors Essence star seed, via pure essence seeding with the soul in contract. Violations of genetic free will, body hopping, stealing energy, or energy holding were unnecessary. Essence IAM is its own natural life wave carrier throughout the cosmos. However, key in the new species life wave shift is the final bio-embodiment process, where the new Q heart must merge **_both Male RNA and Female DNA Emotions_**. Here, the old unconscious polarity emotion human heart merges into the new

Essence light heart. As we review these ancient matrixes you *can check to see* if they have resolved and re-merged, as One New Divine Male-Female liquid light flame in your own Cosmic Heart. These Creational Parenting Roles that Spawned the Collective Unconscious to evolve into the fully conscious quantum heart have emerged from an ancient creational base of the Father-Mother principles that will be carried forth as new star families in your light universes.

In the **Creational Mother Matrix, the Healthy Mother** is **the first principle of love** and bonding and the progenitor of the DNA in accordance with all life wave codices. She is Core Essence light's right to exist. She does not kill unless in an aberrant state. She cycles all emotions through the atoms of: earth, air, fire, water, and ether, as well as their newly born quantum essence particle light. She represents: home, hearth, children, bonding to cell life, Essence love, relationship, and partnership. She must express her core light potentials as feelings or creativity, and will feel a full range of positive and negative emotions merging them into Divine emotion. She knows that her love protects and allows all emotions for evolutionary growth through the heart. She makes all experiences safe, protecting the right to exist under free will. She is the heart of the imagination and her desires are raw, primordial, instinctual and coded by innocent symmetry. Her will is natural free will. The anti-matter void is her womb space and the birthplace of her creativity and the focus of her love until an enhanced new quantum heart births conscious dark matter or particle liquid light creations! She honors her beloved male twin self to sustain all that they create in all forms of life they choose to create.

The **Aberrant Mother** focus is on *dark and heavy emotions,* which replace sisterhood and soul bonding. She will abandon, abuse, and use her children for her own needs. She will compete with other females for power. She even poses as the dark primordial abyss that devours males and can shape shift into any of the twelve archetypes: warrior, priestess, Madonna, Medusa, teacher, etc. She is subject to the negative altered-ego as the collective unconscious where pain, shame, blame, judgment, death and suffering predominate. She destroys in violence rather than recycle in the natural order. Violence replaces natural cycles as orgasms, volcanoes, and birth becomes violent. Sexual fantasy replaces imagination and the Divine ecstatic bliss bodies. Physical, sexual, mental, and emotional enmeshment replace bonding. She feels not enough as a woman and will hide in the Void, sometimes withdrawing the life force and going into the nothingness or further into Black Hole. She will go into attack and defend as a standard defense. Here she questions whether or not she can hold her own love. She feels banished, parched, and forced into the underworld in sacrifice. Her war with the male is always over the fact that he took prime RNA patriarchal role as Creator and in the creational merge. She struggles to understand the very male emotion she has agreed to carry in the Divine DNA belly of her creations. Sacrifice has replaced sacred living; fear and death have replaced imagination and Divine desire. Worst of all, she feels the pressure of the male helix RNA to gain control of all the genetic bloodlines, which by Genesis, is her natural function; whereas he only transcribes the gene codes. She experiences the projections of male anger, rage, and violence for the first time and feels abandoned, betrayed, and unsupported and in competition with her twin male-self. Yet she needs his male RNA genetics in order to sustain her body. In reality, her emotional reactions of judgment, and fear created a foreign ruler or inner terrorist.

In the **_Creational Father Matrix,_** the **_Healthy Father principal nurtures_**, supports, and is in service to the female creations through _holding of the outer light._ His nature is the creation of: the Medtronic atom, manifestation, time, form building, architecture, movement, protection, maintenance, problem solving, boundaries, mind, and the preservation of life. All actions and experiences are sacred; and no experience is better than another. Forgiveness is inherent within all movement. His domain is the external reality and its forms, structures, and functions. His will is that of the Divine Father Source codes. His male heart is toned by his RNA genetics which transcribe the mother's bloodline's DNA's Love codes to support and self-sustain life. His equality is in perpetual support of the mother creations and in manifesting her love focus. The ***Aberrant Father*** **emotions are locked** in the negative male 12 archetypes such as dark priest, star warrior, magician, abuser, womanizer, etc. He sees the female as weak but still having creational matriarchal power that can challenge his power. His emotional reactions are expressed in abuse, rage, anger, violence, control, blame, shame, etc. as a defense against the creative power of the female, which he denies as his twin self. He will kill her or their children unless they fit into his power forms or mind controls in his supreme right to rule. He lives in the mind and outside his emotions and expects the female to carry their expression for him. He greatly fears the Void and that he will be castrated in the Great Abyss and cease to be. He competes and wars with other creator gods for control of the universe. He is subject to the negative altered-ego and collective unconscious beliefs. Power without love postures in a false ego mind-emotion that fosters pure male will without adaptation or change, even to the point of cutting off or cloning his Essence Divine DNA blueprint. Here, creating outside core light becomes more important than loving and sharing. He will force the mother, the other half of himself, to in-volute into Black Hole and lose evolution or hide in the Void if absolute control is needed. Or, he will create the illusion that he has killed her. His anger, rage, violence, and fear predominate as male emotions, rather than allowing any vulnerability or his wounded heart to feel true emotions. His mind mimics emotions as a substitute for ownership as love. He must keep the mother as second principle of creation to remain in control. This could be called reverse polarization. Mind can't feel but must mimic feeling replacing any divine love with a condition of power. So, addictive mind postures as love. Love is often a synthetic experience of technology as an avoidance of the experience of true direct emotions. Love in the mind is accommodation, assimilation, or obliteration.

*However, within their Divine Matrix is a fail-safe code that allows the DNA Essence Mother to re-create the Essence RNA Father and vice versa, if there is an attack or threat to their Source Core.* Hence, your bio-mastery is consummated when 144,000 density-dimensional/s, or dark male-female aspects and light male-female aspects have all remerged, sharing, releasing and re-uniting as all experience. Master Self then lives inside conscious Love always. New heart in Divine marriage is ready to bring relationship into a quantum age in multiple helixes of expression, with all new roles to play in all new creation stories, as potential expressions. *You have arrived in Love's Essence, over and over. You have pioneered the courage to keep turning self-inside out. You discovered that Self Love demands ALL of You. The prize is mastery over the heart; for that is the quantum Essence body, where love is inclusive of all life! Re-emerging as your own core light is never again masked by the mind, by the human, by power, and the seduction of death, disease, and suffering identities. However, <u>no organic bio-essence soul escapes this process.</u> There is only one question asked by the child-the soul-the spirit on the last day of time by the cosmos? DO you love who you have become? Finally, new heart flame realizes that living in the Presence of unique Essence is Love's greatest expressive potential. To bask in new Love, be Loved, and live in the Joy of the hew Heart in those sweet unexplored places of new consciousness, illuminates everyone and all life! How wondrous it is to release programmed love; to have no attachments, no holding back love, no hiding or controls over love; or the human, trying to be better than those it loves! But now, to be always bathed in the One Source Essence flame and to have any experience and still love Self and Other.*

# THE NEW EARTH

### THE SEQUEL TO THE STORY OF LOVE AND CREATION

Aieou

Aum

Self Mastery

## MAURENE WATSON

# Perfect Creation at the Multi-Verse Light Summit 5-2020

*Q: How do I know if I'm following my new life's path?*

Masters, **Creation is perfect in Evolution's Life Master Energy Codes.** Most of you hosted and presented transmissions at the **Multi-Verse Light Summit in January 2020** on **Venus**; where you shared state of the art progress across the cosmos of your new standard of consciousness. You also got a transponder video transmission from the ALL in ALL consciousness of the astounding potentials from the orb consciousness-Nautilus-Life Spiral. You viewed your New Earth's, planets, solar systems, and universes that have been created out of Old Earth's wisdom School-Universe. Your planet is now moving into unprecedented quantum acceleration. Your now aware that you were here to fulfill potentials both in the world of atomic matter and its ascension into the worlds of quantum or dark matter potentials. This new standard of consciousness offers a new species master DNA life code to humanity as they enter the Golden Age of Aquarian light in 2024 and seed new Earth colonies in your local universe; and as the seed for the multi-light universes in becoming. However, they will **use the technology of consciousness to awaken and reconnect with their star families; now that ascended bio-consciousness has been anchored.** This is so they understand how to create with quantum matter, which was not mastered in the age of Atlantis timeline and caused the last demise of Earth's crystal transmission life network, through their misunderstanding of free energy. Humanity will again use technology to expedite their heart's core light connection with their star families and to remember that their original life code was to imprint themselves as a new Species. This gets **transmitted into new** technologies, innovations, starships, business, commerce, and light body understandings of heart relatedness. Such awareness will help you integrate and navigate the essence embodiment of the new life cycle you are in as you become what you create. It will also aware you that that **receiving your own creations** and accepting all that the cosmos freely offers is real giving; because what you are willing to receive goes throughout the cosmos multiplied-quantum. Hence, there is no greater service than receiving such love. It was also quite the event to celebrate your world's enlightenment as first new Soul Essence DNA representatives of your new human species at this Multi-Verse Light Summit.

Your now aware that **evolution follows its energy life codes perfectly,** and it is time for you to do the same, as you experience yourself more and more as free energy that serves all life. Your entire planet must now also take responsibility for their own energy as well. You will also aware yourselves that everything you do in each moment is Perfect for your own unique creation as you explore unknown potentials. Your new heart essence master codes inside your sovereign star-sun deem it so, as the inner Earth's sun opens its plasma core into a magnetic-quantum transponder, discharging your soul from the old atomic electrical field. This causes the mind-emotions and human biology to unravel, into death, disease, or suffering, if heart field can't stabilize the soul core light. Such collapse of the old electric Earth field and all its time lines was seen as Armageddon, but is truly migration into crystal-diamond-particle-plasma light fields and their potentials for the New Earth universes. Now your aware that your free energy vessel and that of Gaia are in a parallel process. And you realize that everything comes and goes from consciousness, which individuates as essence expression creation, inside life code existence in the physical and non-physical realms.

Masters, any communication or talk of being afraid of technology, aliens, wars or anything outside your own energy bio-sphere is a distortion. If you're truly a master of your soul's core awakened bio-light; then your Heart's new bio-Essence imprint changes your reality moment to moment. Your 6th sensor; or your pineal orb diamond-star, transmits the visons of your heart's imagination on your frontal lobe light sensors as your heart essence experiences them. The outside world responds to your illumination. You do not react to any energies outside you. There is/was no enemy, except your own inner aspects who perceived inner darkness or polarizations as outside reflections of self or against life. Darkness is just the absorption of the creation womb light spectrum reflection. That is why you called upon the IAM Presence decree of the violate flame of: ultra-violet, gamma and cosmic rays to activate your full light spectrum, (both quantum and atomic), to release any energy that did not produce perfection for you! Let's example evolution's life codes in your Old Earth: local, linear, or holographic universe; and how their perfect symmetry guided the heart of creation and grew your soul's new life's Essence codes. Masters in the Old Earth Universe, you had been under such extreme distortion of life code that your organic joy and all its potentials had been masked. You are now in your *** paragenetic code vessel, hosting full spectrum Homo Sapiens essence spirit light and not the old human genome epigenetic body, which just hosted human DNA atomic light. These new Life master energy codes of blended atomic and quantum rainbow frequencies transfer star information and star matter for ascension in your new life cycle of evolution as enlightened creators for your new species human race. This was the star-seed foundation Being for the Cosmic Creator in emanation, infusion, and incarnation of World Systems for the Divine IAM Presence; or Source in form.

Life codes master energies in your old Earth Adam-Kadmun, or your atomic human-DNA heart. They were also stored in the: crystal skulls, arks or rays of physical plasma light; in your planet's inner earth chambers and inner core sun, all earth crystals, and all living organic elements and cells of the 4 kingdoms. This included whales, dolphins, jelly fish, flowers; and vault chambers of illumined master embodies from the cosmos who took human incarnations to re-seed lost codes. There were the Solar System Lords of Light embodied as 12 Enochian Master energies from each of the 12-constellation house seedings for Earth incarnations star material. Each carried the 12 patriarchal-DNA bloodlines, or transcription codes for descending worlds, to procure humanity's full consciousness including upgrades. Hermes Trismegistus, (https://en.wikipedia.org/wiki/Hermes_Trismegistus), sponsored 12 Emerald tablets or master-energies, acting as radionic energy fields of Living light. These acted as an organic computer template of symbols, which radiated the Seraphic Flames of the Elohim programs/blueprints of Light mathematics. They coded the sacred geometries of DNA encodings into physical humanoid life for your local universe. One of its texts was the ***Book of the Secret of Creation and the Art of Nature.*** These light fields later manifested as rays or arks, (archangels), of light or dimensions. The crystal skulls were the 72 Metatron crystalline atomic cell light codes for the human race. Your Earth pyramids and ancestral mounds operated as electro-magnetic generators, interstellar light networks, and landing ports to communicate and transport with your star families.

Recorded in the **Book of the Tau,** the tetragrammaton, (https://en.wikipedia.org/wiki/Tetragrammaton), or master energy imprints within the radionic templates from Orion 13,000 years ago, re-calibrated distortions in the magnetic charge of the planet by re-seeding the inner earth races and the God-IAM-code memory. This re-ignited electro-magnetic crystalline earth networks with the magnetic charge of the living radiation of the IAMTHATIAM, or the lost word of Abrahamic religions and their quest to remember their Divine-DNA origins and upgrade civilizations of humanity. This would embody within linear time, that Earth's inner sun or future light body would create high-frequency rotational fields of plasma around its core too align with all cosmic suns in new Earth's future potentials for enlightenment. This was to also counterbalance any distortions in the creational energies. This would also result in igniting Earth's mother-ship core warp drive when she goes super-nova as her own New Earth-Star Sun, just as you will do in your own free energy bio-vessels when you leave Earth.

Just 2,000 years ago, came an attempt seed embodiment of the GOD IAM, to new levels of bio-awareness. This essence expansion of the heart consciousness and its qualities came; via the Master I'shoa (Jeshua), at his baptism in Jerusalem. He seeded the Cosmic Archangel Ray ship or bio-vessel of light in the Essence Homo Sapiens Being. And, as a Collective Jeshua of Star Seeds, most of you also embodied in the human flesh to seed the light vessel codes in the Heart of humanity's New Earth Star future and the resolution of its Old Earth's destiny. I'shoa's "home" was a planet of the Blue Star Rigel in Orion. Its distorted DNA antithesis was/is the star Betelgeuse. (**https:// enwikipedia.org/wiki/Betelgeuse**). These codes were imbued with such androgynous essence qualities as: star-walker, angelic messenger, time-space traveler, compassion coder, pillar of light, new worlds directive, unknown manifesting free will agency, or heart's creations. Those of you masters, now in your enlightenment, were in or had soul aspects in and out of lifetimes; to provide collective support for this and most planetary upgrades.

And now, these master energy Life codes have been upgraded and transferred to the new species Homo Sapiens free energy vessel of new star matter that each Master has prototyped. Each is its own bio-sphere network of light within its own new star-sun. Your core Heart imprints the 3$^{rd}$-eye in each moment to trans-sense new qualities of experience in your new essence imagination worlds. You will now host quantum interactions at multiples beyond the speed of light, color, and sound, bridging all physical and nonphysical new Heart master life codes. These regenerative light fusion fractal codes function in perfect symmetry like the progressive infinite iterations of the "Nautilus" that you saw at the Light Summit and that appears like a seashell or the spiral of life, of the Mandelbrot Set, (https://en.wikipedia.org/wiki/Mandelbrotset) and have operated as crystalline logarithms of life. Humanity will use these codes in their new technologies, to innovate and upgrade every life code of their bio-light life systems, with their New Earth star families and the New Earth colonies that are being seeded.

So, in your perfect moments **of perfect creation**, you simply allow the Eternal Heart Presence of your illumined Eternal flame's unique spirit, to vibrate your HEART's essence imagination, to answer to these new energy code potentials, both physical and non-physical. Heart hums or pulses, vibrating your composite soul's unique core light's__ crystal-diamond-particle plasma essence flame____ as the warp drive in the heart of your mother-ship. This vault or Holy Spirit container allows you to transport into quantum realms, imprint new essence matter, and enjoy new human sense essence. All the while, you're having a cup of coffee and animating the Divine Human till you no longer animate your imprint on this Earth. Enjoy the perfection of your every moment now as your Heart code guides you until you are fulfilled by all New Earth's potentials. And yes, **every moment you are in is perfect for you!** We also reiterate that; your new living codes guarantee fulfillment because you understand the bio-organic consciousness and energy relationship. Your now aware that Life's Presence is in an energy relationship which is alive, passionate, expressive, sensual and full of intimate communication. Consciousness is sense-essence energy expression including all the qualities grown by the soul's heart love. Free Energy lives in the Presence of all Life's existence. Everything is energy, and energy is everywhere. Life's heart flame is *passion in motion.* **The new spirit heart is the soul's energy imprint.** All Life animates its life code and you are in total receivership of the ALLINALL, simply by just illuminating your light Presence in the Essence expression of free energy. Heart vibrates information of consciousness inside your new network light codes in your new DNA that excite your Essence perfection in your new qualities of love. Indeed, they were grown from the **Homo Sapiens gene of compassion**, which kept giving and receiving equal and harm to self or another impossible. Service to self and service to other have merged into a new species. ✳✳✳ **Paragenetic-term channeled by** Maia Nartoomid; **reference: newearthstar.org or spiritmythos.org.**

# NEW EARTH
# LIGHT BODY

## MAURENE WATSON

# Potentials of Viral Energies 6-2020

*What is the higher purpose of these viral energies?*

Quantum Masters, we must reiterate in **channeling You back to You**; that there is much apprehension and excitement about the quantum changes that your light has illuminated to the surface of your world. First, the cosmos thanks you for your service to all humanity Light Masters. Thank you, for sharing your enlightenment and your Master of Ascended Potential with: humanity, your new worlds and the Cosmos! Thank you for being the light, the love, the prayer, the miracle, and the magic. Thank you for being the embodied daystars that you are!

Indeed, you have all seen the effects of your new DNA consciousness as the inoculant for any: unnatural, distorted, mis qualified energies and that multi-diversity is your planet's greatest strength. **Indeed, <u>your planetary reset </u>is underway in: economics, cyber-ware, bio-astro organics, transponder technology, meta technologies, and star-sun resets with all the New Earth exoplanets. Included as well, are changes within the new Homosapien light bodies; along with understanding energy essence heart relatedness, and it process in creating. Indeed, Preparation is already underway to move from global to multiverse citizenship. And indeed, you will also become the antidote for the next viral energy serving humanity's choice between becoming a global heart light network of bio-organic life or just a global technological mind.**

Your aware that everything is energy. And energy is free as in free will. Hence, **YOU, as your universe, have all created an energy, an organism of consciousness that serves to equalize all your choices such that the entire cosmos and Its consciousness may fully embody and experience its own Light and take it into the next stage of evolution.** This viral organism is in service to all life and answers to the individual frequency of each consciousness. We will make a **list of some of these offerings of higher purpose** and you can add your own. You will realize they are endless and **multi-frequency variant** as the emerging light exposes all hidden agendas, as well as outcomes of new life's transitions.

This energy offers a time to end **the matrix and Old Earth energies** for souls who want to be free to create from their hearts and not bound to external energies of manipulation, control, or power. In this is the release of planetary collective unconscious fear that has been trapped in your creation for eons. What **brands of war** will humanity continue? Will it be: military, cyber, drug, surveillance, viral, nano-technical, economic, religious, racial, genetic, or in space? Or will they realize that the enemy is always within? Will they realize that their collective: thoughts, feelings, attitudes, and beliefs create the outcomes of their every moment including their weather, their health, and their governing structures? With power brokers and experts no longer **the heroes of hierarchies**; they can no longer feed off the energies of those undervalued souls who either serve humanity with tender hearts or those who toil like slaves to survive. Your planet has always thrived on the energy of glamor to keep its cast system of value alive as in: Hollywood, billionaires, soldiers, sexualizing gender roles, sports warriors, and now its worker bees. The hero worker bees that keep the economy running are each telling their stories in the media like movie stars or authors of their life's book. Indeed, traditional Earth School is leaving behind a legacy that will forever be renowned for changing the **divine right to life's existence in equality** for the entire cosmos. **Will each soul become the hero and small business of their own life?**

This energy opportunity offers humanity a time to realize that each soul is responsible for their own energy as their own creator; and that everything is within them, even **their <u>own vaccine</u>.** They may be shocked to find that they are the only acting authority in their life; and if they attend and care for themselves, then the love will overflow onto others by transmission, if not by touch. And some may realize that in the end only love can save them because the self is the other, yet individual in its own soul's purpose.

Humanity was **<u>stuck in their own energy</u>** and will no longer be able to throw their negative emotions and trapped energies onto another. Instead, they will find these energies trapped in their own chaotic mind control, sick emotions, or a worn our cell DNA fragments. If they are in a trapped polarized fight with their own energy; then they will need a vaccine. IF they are accepting the voice and true tone vibration of their heart, then their core light will serve as a regeneration chamber and their heart will inoculate them; and attract exactly what their life codes need to continue on in their soul light to grow their soul's fulfillment. **Science, technology, and medicine will find all its answers inside the <u>blueprints of nature and the human biology</u>. This astrobiology will help them discover that they all imprints of the stars and suns in the cosmos.** They will discover all that they need is already within each soul.

This energy is also assisting many <u>souls to transition</u> who are ready to step into their next life. IT is teaching many about what is mercy? What is the value of a life? What is the value of a body? Is the body a heart vessel of light or a sack of just human chemicals? What is the *value of the soul and where is it?* Is how you treat the least of you really a measure of the soul character? What are the qualities of compassion in equality and respect for service to self and other ass a measure of do no harm? What is a measure of individuality within oneness then?

This <u>energy in service to all life</u> and all variant soul frequencies, allows a world in quiet quarantine so humanity can hear what their soul's purpose is, as they disconnect the programmed cage of external demands and follow their own heart code instead of a herd mind. Humanity is sensing a new master Presence on the planet that whispers in their ear that says there is more to life than empty labor. They may sense that if they can just awaken to their heart's core essence light, they can heal themselves and that will heal their planet. This energy brings up the Divine **right to exist** in the underbelly shadow of multi-diverse inequality in dinosaur institutions and *outworn evolutionary human systems.* **These buried shadows <u>of throw away souls</u> is reported in your media as: prisons, refugee camps, indigent medical facilities, homeless cities, gender and human trafficking cities, DNA lab factories, religious communities, mind control cities, and general in-house domestic violence. All remain pervasive due to the abuse of the heart's DNA-Essence bonding code.**

So, this energy offers **queries?  Was** Earth just a prison for wayward souls? Is there a god, a creator, or a hierarchy of angels, aliens, and star gods who Earth answers to? **Who does humanity pray to, channel, and who ghosts them or abducts them from their bed?** And if there is a heaven; where is it? Will they ever realize that everything is within their own consciousness and each is their own god within the essence heart of their own creation? They ponder who is running creation; the nonphysical realms or the realms of matter density?

Humanity is learning that in evolution, energy of any kind cannot be trapped, bound, powered, manipulated, controlled, or stolen. **Energy is free** and consciousness gives you access to as much as you want to play with. Consciousness, energy, and matter, and *their relationship* is what humanity came to master through the marriage of all the nonphysical and the physical realms in your local universe. These are atomic and quantum energies of the light spectrum: hertz, infrared, visible light, X-ray, ultra-violet, gamma ray, and cosmic ray. **And the ultraviolet frequency is the <u>soul's inoculant</u> for all life's bio-organisms. The soul's overall mass to light frequency ratio dictates** how they receive the consciousness of any viral energy. Will the awakened master soul codes of humanity short circuit any illusions that it has to give its privacy and civil rights away to the medical-pharma cartels who make trillions to patent DNA, viruses and vaccines; in order to subvert New Earth's Enlightenment? Will they realize the risks of surveillance, nano-viruses, and artificial intelligence choices inside freedom or invoke their soul consciousness to use technology to serve all souls to thrive and prosper in more efficient and creative innovations. Humanity is also exploring all the nuances of compassion using these frequencies as spirit essence, angelic senses, and human emotions of both extreme light and extremes dark qualities. Do they want pink saffron black jelly hugs or do they want rainbow peach aqua golden white light kisses? They are exploring what is their heart and who should live it with them; and if they can trust that soul heart will truly guide them rather that the chaos of the mind. Do they want denser body sensations or senses of light; or perhaps a blend of divine human? Or, will they decide to be a techno-human? With climate cycles they must query the **value of nature,** time, space, gravity and its connection to the soul and what place does **quantum technology offer in that <u>relationship</u>**? What happens to energy when you force evolution or try to alter the natural life code of any species? Is the virtual world of technology compatible with nature? Does humanity or their planet regenerate or destroy is own species? And is there room for marriage and family in a supra modern age of a transhuman? Is there room for real intimacy and love or will it be just supra glamour of a space force life. Will children be born: in the womb, invitro, via soul transfer in a light body? Does creation regenerate life via its soul essence DNA or via nano-cloning or a hybrid combo?

*Therefore, eventually, **each soul in its own composite frequency, as it leaves the Old Earth Matrix dream, will eventually have to release their old pasts and futures just as you did. That includes: their old angelic families, old time lines, old DNA, old molecular past bodies of death, disease, and suffering just as you did as a living example.** But it is also your time to come into your true master self and enjoy your own self realized core essence light as you begin your ascent into the New Earth realms, while fulfilling and unveiling new potentials in the process. Masters, celebrate that as graduates of the Old Earth Universe, you move into the TRUE composite frequency of your individual soul's evolution setting the new standard of consciousness for the new DNA heart essence species of Divine Human and Potentials for all the New Earth Super Universes of light to come. Enjoy and celebrate your **graduation into enlightenment and your Master of Ascended Potential evolutions as they spawn new star material across the cosmos!**

# THE FREE
## ENERGY VESSEL

# MAURENE WATSON

# Planetary Reset and Diversity of Awakenings for New Worlds 7-2020

*Q: Can you please review how love's bio-cell bonding work in the multi-diversity of the light vessel?*

Quantum Masters, there is much apprehension and excitement about the quantum changes that your light has illuminated to the surface of your world for the planetary reset and its diversity of awakenings for the New Earth World**s.** You have front row embodied seats for the greatest shift that has ever occurred in the cosmos, with your Earth as the flash point, for a new bio-human species in a vessel of light. Your old Earth genetic Universe has yielded the effects of the New DNA consciousness as the inoculant for any: unnatural, distorted, mis qualified energies and that multi-diversity is your planet's greatest strength. Indeed, your planetary reset is underway in: economics, cyber-ware, bio-astro organics, transponder technology, meta technologies, and star-sun resets with all the New Earth exoplanets. Included as well, are changes within the new Homosapien light bodies; along with understanding energy essence heart relatedness, and it process in creating. Indeed, Preparation is already underway to move from global to multiverse citizenship. And indeed, you will also become the <u>antidote for the next viral consciousness,</u> which will be humanity's choice between becoming a global heart light network of bio-organic life or just a global technological mind. Your beautiful essence heart codes are still meant to touch, hug, and bond and not be separate from each other as Divine Humans in heightened sensations of light, color, and sound. ***There will be no more programmed <u>syndromes of 'skin hunger'</u> that your last virus quarantine revealed as an anti-life and angelic-organic essence sense. The agency*** *of sovereign bonded love has grown into your multi-diversity and that is the strength of your new species race.* Your new rainbows of luminescent skin have already embodied the density of the atom and this is carried into love's quark particles of bonded essence senses. Your virus has spoken and the days of the Old Earth being a predatory food chain and evolutionary toilet are over. It will again speak your choice potentials about artificial intelligence or the bonding of the New Earth's bio-organic heart Life!

Hence, it is important to again remind you that those soul's just awakening, accepting enlightenment or ascending will *fulfill their soul's contracts in many different ways.* As you know, Old Earth's wisdom School-Universe and, your planet as a prototype for all the New Earth's and the coming Super-universes of light, is now moving into unprecedented quantum acceleration. Your now aware that you were here to fulfill potentials both in the world of atomic matter and its ascension into the worlds of quantum of light matter potentials. You also know that this new standard of consciousness offers a new species master DNA life code to humanity as they enter the Golden Age of Aquarian light and seed new Earth colonies in your local universe; and become the NEW Earth's for the multi-light universes in becoming. Beginning communication and contact with your multi-light New Earth Star-suns, that is scheduled for 2025, will now be availed to each of you individually as your Composite soul vibration chooses for you! Remember these new worlds are first within your own consciousness. That is soul's purpose for preparing your light vessels and activating your new DNA heart's life codes, so you can receive such transmissions and fulfill your choice potentials.

You have gone the way of spiritual warriors via creating a new species bio-organic consciousness to prepare the way. However, humanity will use the technology of consciousness to awaken, release their old star families and reconnect with their New Earth-stars; now that ascended bio-consciousness has been anchored. *These new Earth-stars are not families, but will embody light beings of composite equal resonance continuing to explore and complete their SELF realization or enlightenment.* They can choose to create and explore light matter creation in any way they can imagination in new essence qualities of joy, passions, playfulness, fulfillment within new octaves of conscious vibration in moment-to-moment access. Here, Creation opens to a level of intimate trans-essence experience never experienced before, because the light vessel is the base bio-physics. They will not and are not bound to create group star families, as in the distorted-polarized bonding within codependent trapped energies of the Old Earth; that they have already mastered. *They Presence to enjoy the relationship with their own free energy light core and if they choose, dance with others doing the same.* They are exploring the awareness of living inside and creating inside their own energy as their soul's agency in varying stages of the light body as it continues to evolve into a free energy quantum creator vessel. However, they can still dance or share the expression of light essence with others as a bond of life because they are first bonded to their own light. This gives the freedom to understand how to create with quantum light matter, which was not mastered in the age of Atlantis timeline and caused the last demise of Old Earth's crystal transmission life network, through misunderstanding of free energy.

Also, remember that your new consciousness gets transmitted for all Old and New Earth humanity into: new technologies, innovations, starships, cyber-business, individual commerce, and light body understandings of heart relatedness. Look to what science is calling astrobiology to reveal new life and its organisms in the cosmos, which you have already seeded in your new light body vessel heart's master codes. Such awareness will help you integrate and navigate the essence embodiment of the new life cycle you are in as you become what you create. It will also aware you that that receiving your own creations and accepting all that the cosmos freely offers is real giving; because what you are *willing to receive goes throughout the cosmos multiplied-quantum.*

So, please **_honor all the ways_** in which humanity raises their soul's vibrations and migrates to their next lifetime or existence. That is why there are so many of you sharing your inner journey's wisdom template as living examples. You have shared your stories as: inventors, teachers, guides, mentors, channels; and in websites, books, light groups and new tools you have all made available to assist and play in. Your energy will attract those that will have their heart light's creativity stimulated as they reflect off your living example and your illumination as to what creative gifts avail their own fulfillment. This is such that, even another may find the core light energy within their own Divine Presence. And if you do not wish to share your own creations, simply illuminate your light and joy, as it also transmits to the whole of your worlds. Your also now aware that **evolution can't be forced and follows its energy life codes perfectly,** as you experience yourself more and more as free energy that serves all life. Indeed, everything you do in each moment is perfect for your own unique creations, as you explore unknown potentials. The awakened soul codes of humanity, which can regenerate and resplice their own DNA will short circuit any illusions that it has to give its privacy away to the medical-pharma cartels who once made trillions to patent DNA, viruses and vaccines; in order to stop New Earth's Enlightenment.

Again, the new free energy inside the *relationship of consciousness, energy, and matter* will reveal itself to humanity now in: astrobiology sciences, new materials, new species, new foods, multi-resource energy systems, new ways of Earth and space travel along with magnetic transponder-energy technologies, cyber-currency exchanges, self and global governance, biology regenerative/s, light body education tools, and housing. Also, new models of marriage, family, and healthy self-care, will all bring an equitable distribution of value, resources, and energy across your world. **<u>Life transitions/migrations</u>** will replace outmoded beliefs in death, disease, and suffering as **natural light vessel bio-*evolution*.** And even the way children are incepted and birthed will stretch new concepts in essence regeneration, genetic editing versus hybridization or cloning. And let's not forget the beautiful new organic super-conscious divine-human senses; that have grown the Essence Homo-Sapien gene of compassion into trans-essence skills and abilities; and have integrated into your new DNA life codes, since you were in the angelic realms. These embodied angelic human gifts will/have created the light body standards for the new Divine Human species across your new worlds as it grows into multiple bio-helixes, yet to aware themselves to you, throughout your Super-Universe. Enjoy the trans-senses and smells of a new quality of freedom the cosmos has never known. Indeed, enjoy your multiverse soul-purpose and new life potentials for every being on your planet only to be seeded, matched, or assisted by your long awaiting New Earths. Yes, whatever you can imagine, you can create inside the **Essence-multiplicity of new hearts of multi-genetic diversity.**

# Compassion for Humanity's Grief 8-2020

*Q: How does a viral energy open the heart?*

Quantum Light masters, your aware that with one drop of water you can create all life. Indeed, one tear shed from the heart can change the entire cosmos! Your **Heart's** new DNA spirit super-conscious or trans-senses can feel new life in every drop of the ocean's liquid light. Hence, in the final stages of the awakening offerings of the 2020 viral sequences; grief has an opportunity to open the global heart. Your also aware before humanity moves on to the choices for their next Bio-cycle of evolution; they must now grieve their Old Earth creation story just as you had to. They must grief any mis-qualified energies allowing themselves to be seduced by its fear, control, and mental programming; as if creation and energy itself could be forced or hijacked. It was a journey into existence, not a forever identity. They must mourn its distorted energies, false dreams, and addictions to suffering, death, and disease they have been under the hypnotic illusion of. They have to mourn the loss of their connection to their soul's-spirits light **and organic essence bonding code of its DNA stem bio-love**. They have to mourn replacing receiving all that life has to offer, with giving in suffering. They have to mourn the loss of those that they hope will have the promise of life, they have yet to live, as they migrate to their next existence. They mourn that they have let the mind torture them. They have to mourn not listening inside their hearts to their core essence Presence whispering in their ear that it will caretake, nurture, and love them if they embody life's existence, including no need to validate or justify it over and over. They are grieving the that they gave their love and worth away to the viral energies of power agendas and controllers, who would barter their right to exist by selling them fear. It would come as the need to **validate** *t*heir equal right to the unlimited resources of creation offered by the freedom of their own creations. More than any time in the history of your universe humanity is feeling everything that it created. And its **grief is simply the underbelly** of its **violence, anger, blame, shame, hurt, abandonment, doubt, judgment, regret** that were never resolved. These were the trapped energies in your universe of spirit essence and angelic senses that later densified into negative human emotions.

Now, *your compassion understands such grief.* Not only did you have to become masters of limitation, but also masters of illusion. Many of you as you come into your full light have and still may be grieving the loss of: the old safe predictable life, the old lifetime identities, life times, wounded patterns, old spirit families, old limitations. You grieve what you knew as safety from fear. It was the past suffering of a distorted creational story where competition and separation became suffering rather than the gifts of receiving all that creation has to offer. You came to accept limitation of every form of fear as a prison of trapped distorted energies you would never escape. But now you know your grief is your own and allows the DNA bio-cell transformation of any core pattern to dissolve and re-imprint your new life codes, such that false security is no longer the death of bio-genius. So, allow the grieving. There are many levels, and stages of grieving and each soul does so in different ways and for different lengths of time. You know that your tears cleanse the wounded core pattern of: the right to exist without having to validate yourself, disappear, hide your light, prove worth-value, pretend; or empath for other's suffering in the constant dying of the old body, wounded lifetimes, fragmented DNA codes, and past-futures. Indeed, you all know what it is like **being on the edge of the final rebirthing** of your core essence light just as you are about to awaken this Heart being that is the real you behind the mask. It is truly uncomfortable and can be miserable as the mind tortures you. You cry to the heart of your spirit: *"Please don't let this mind game and its nasty limitations blow me up, make me insane, or keep the polarity arguments torturing me like a robot, before I manifest/self-realize my dreams. And maybe I might even meet my true self in a burst of light?"* You bounce off polarities trying to ground your spirit's fully embodied light. Then you realize real abundance, **real manifestation is self-value, self-acceptance, and self-trust of your spirit** who has indwelled in a vessel of light you created for it by mastering all human experience and a dark night of the soul. And the spirit is so happy to review all the growth, movies, and wisdom since it was separated from its soul and human aspects. *And, you have even grown a new heart biology of light to house it.* Even trusting your Spirit's Presence remains a stretch as you fluctuate your multi-variant frequencies inside the physical and non-physical. And just when you felt you fully integrated your light body another aspect in the higher old earth realms of light who had never experienced density needed share its movies to be bio-merged.

So, dissolving all the bouncy polarities is what you should be feeling because you **were never fully embodied before.** And now you feel any distorted memory pattern in every cell as it cries and dissolves; so that in the new vessel, Spirit's love in the cells will be bonded in the DNA/ RNA master code. Finally, you are *bonded* to your own creation. Bonded in a composite true stable light quotient or frequency of your own Soul and Spirit IAM, while embodied in a new vessel or quantum instrument. Here you know the result is that the Quantum heart is its own: regeneration chamber, imprinter of new matter, a new master code DNA species life form, and its own bio-sphere star-ship. Hence greatest **manifestation at this stage is the Trust** of your own inner connection and the inner bond to your Heart's excitement or stillness to allow your spirit to reveal each upgrade step by step as the old vessel gradually re-particles into the new model light body.

In its embodying of the bio-instrument that you have prepared for Spirit by releasing the Old Earth Universe, there **is no need <u>for</u> <u>validation of your existence</u> over and over in lost identity, purpose, or lost connection;** for that would make you a <u>**victim of creation itself**</u>. That would just be another distortion. That is why you allowed yourself to experience life review after life review, as you went back and forth to your version of new Earth and reviewed the movies of your life on the old earth simultaneously. This tele-transport in frequencies opened your **true composite essence vibration to remember your first light;** so you don't need to control, power up, or manipulate anyone or anything to allow yourself to exist and receive your new life. And that life is you because you have mastered all life. That's natural evolution. Energy is free and you're the everything of your own creation. You never have to steal or bargain your soul to get it and there is/was ancient ancestral DNA cellular grief there from all prior generations.

*<u>Grounding your light</u> comes by allowing the Heart and all those feelings to arise in your awareness so they are forever: released, dissolved, re-molecule/d, re-essence/ed, re-imprinted, or changed by the new heart code.* It also dissolves the old pain body completely. That's transformation needed to stabilize the wobble of oscillation of the heart's crystal-diamond-liquid light toroidal bubble. Remember you are light and your composite new **spirit heart knows how to transform and perfect every cell with its new master code.** Hence, for all light master's embodiment or master code love in the DNA/RNA cells, is the key and trusting that your spirit in fully bio-souled to rebirth/resurrect yourself out of any distorted wounded illusion patten. This moves you back into a more evolved and composite Heart with a bigger light capacity

and frequency to create. It makes a new space to be filled by your new life's gifts and the potential of fulfillment in their new essence expressions.

Finally, the grief turns into release, freedom, gratitude and tears of joy. So, allow others to experience this process in their own way and remind them to be self-compassionate with their grief. **<u>Manifesting trust is the greatest abundance</u>** of this moment as all else falls away from one's life, such that the inner connection to soul-spirit may be awakened into life's new light purpose and the golden age of the New Earths, no longer hijacked by fear's limitations. **A manifest of trust** appears in this new heart relationship of bonded creation and absolute trust of SELF. Then, the programed core pattern argument moves back into the receptive imagination of the music of your new heart tone; which tunes the crystal bowl of the soul. Let any quarantine of your creations be forever ascended into the joyful knowing that all will be healed as the tears are wiped away. And, there appears a new life that remains to be explored in this new elegant vessel of quantum light as its old grief has become the wisdom of the laughing tears of Life's joy! So indeed, allow humanity the compassion for their grief. You all have lived it well and shown them that their tears will heal their broken hearts and, those they love; as well as heal their planet and the Old earth Universe. Their *salt-crystal tears* will only tell them that the heart of creation lives inside all souls and its heart is the only way in or out. It is the only freedom. *Allow your compassion to tell them; "I honor the enormity of the mourning of your grief of loss beyond measure. It is its own poetry and your own song yet to be written for yourself and those you lost. Thank you for sharing your tears with me. It was quite a moment where I felt everything change. Yes, one of your tears changed all creation.!"*

# The Quantum Instrument 9-2020

*Q: How is your heart code the imprint for all the new technologies that will re-imagine your world?*

The quantum vessel is an instrument of the *'All-That-Is' Presence* in the receptive imagination of spirit heart's master code potentials. This biosphere or instrument houses a new species DNA organism with: antigen immunity, new imprint helixes, and super-sense abilities. It houses a master of, both limitation and illusion, who has also become their own master of embodied sovereign Light. This light chamber container also houses the biophysics of how the heart energy has explored what love means to its embodied soul-spirit. Its sovereign heart energy receives, shares, and communicate its soul's unique genetic master DNA codes with the cosmos. Hence each master conscious vessel becomes its own internal free-energy biosphere. And within it, your embodied master has its own ecosystem, economy, own essence qualities and gifts, as well as regenerative imprints for any life code, as their own creator and as their own reality. Your quantum instrument also serves as **an <u>adaptive imprint</u> for your new species evolutionary organism that is evolving all life** for all the ensouled children of creation. This heart code is the imprint for all the new technologies that will re-imagine your world?

Its simple **Heart Essence DNA code** says: *"IAM which never betrays heart. IAM that creates and communicates only though heart. Only Heart essence can change programming or create new life. Heart Does no harm to self, other, or any part of life. All life evolution follows its natural code which nurtures and cares for all life. Heart is the spirit which guides IAM into new life. IAM Heart that releases all the* *guilt, suffering, sabotage, shame, unworthiness, disease, and deaths of limited self-experiences. IAM that always embraces all experiential aspects of Itself and sends them back into free energy, as their wisdom is integrated. They did after all help me grow my soul' new gifts and meta senses. IAM allows Heart to manifest whatever is required in grace to receive all that life has to offer. IAM living embodied wisdom of the Heart of All that is!".*

Overall, this free energy quantum bio-vessel has become an evolving heart that houses the evolving love of a new species divine human with a new master code soul imprint. These organic consciousness imprints are the innovative technology model **for matter replicators** that will soon change humanity's lives. These **living new matter imprints** will appear in: housing, food, water, clothing, education, and commerce exchange systems as well. Another will be magnetic particle light transponders for both travel and communication replacing airplanes, cars, and satellites. They are akin to particle-plasma beam accelerators that transport or re-imprint quantum particle light. These imprint codes will include **designer technologies** that can upgrade all your planets eco-systems and scan an individual's every need with an individual imprint to replicate any form of matter. This technology will assist humanity to restore their divine right to exist with equal access to all creation's resources to fulfill their soul's journey. It will also serve the needs of the light vessel which requires less density in food and all its requirements. Hence, **different types of human species,** besides the bio-organic enlightened master or Divine Essence human, will also be explored using its consciousness master code imprint for artificial intelligence. These include the hybrid or techno-human and robotic-human forms along with their: life styles, blended families, and bio-imprints that match such techno-human choices. The conscious master remembers aspects of their future selves having already lived versions of these timelines in Atlantis, in their Galactic, and Universe's cyber history. This allows all creators in your universe to understand free will and the responsibilities for the energy of their essence code creations and the effect of such choices; such that Old Earth Universal cycle may come into resolution. It also allows conscious/ new matter technology to offer benevolent and equitable choices for all in the new Aquarian cycle.

So, do celebrate the effects of your embodied enlightenments. Indeed, they are offering transmissions of all these possible outcome potentials humanity is playing out, hosting and mirroring; from an energy **organism of virus that mutates to serve humanity's changing consciousness**. Humanity has chosen this virus as their teaching and awakening tool to understand how to heal their pasts and futures and their ancestral DNA and its lineages. What another beautiful opportunity to heal their entire universe! Indeed, this <u>energy of virus</u> offers unprecedented potentials in answering to the composite vibration of each soul as to their next step such as: upgrading their outmoded dinosaur systems, knowing and spending quality time with themselves and human family; and DNA cell healing of negative thoughts, feelings, attitudes, and beliefs. Its energy offers the asking of their soul's core light purpose, healing their biology, creating new careers, listening to their heart light, or even crossing through the death realms. Your virus has answered you and said; 'that you have chosen that the days of the Old Earth being used as a host to predatory energy is over and the divine right to equal access to life's natural resourceful abundance is available to all humanity in equity. It will again speak choice potentials about using artificial neural link intelligence or conscious technology to support living as a bonded essence of the New Earth's bio-organic heart Life. Creation's energy if free and serves all life's choices.

Virus and bacteria are natural evolutionary organisms that evolve all life and provide natural Essence DNA as its own antigen or immunity. However, the new species DNA soul heart is not a virus, but a new **imprint biologic life code and eco-system.** As such it provides a finely tuned: new life container, heart chamber, crystal bowl as an instrument to house its new soul star material. This now comes from the continued ascending vibration of the expanding soul's core light body in new awareness and new experiences. This life code imprint heart hum is the soul's song in the instrument's soul heart's composite vibration. This vibration came about as your female essence soul-star heart DNA and your Male soul-star heart RNA merged and birthed a new species quantum star-sun heart cell imprint which is you! This heart corona or biosphere spins at the unique vibration of each new essence embodied spirit. These twin hearts merged by mastering the atom's time, space, and polarity matter creation to raise matter into quantum vibrations. They married all their past and future selves into a new heart-child of spirit which is you. This new vessel or quantum heart instrument is **like a crystal bowl** sealed in the corona of its newly born inner star-sun biosphere that vibrates to the heart of all creation. So, the birds know where to fly, the dolphins follow the sonar of the sea, the sea flows to the magnetism of the moons; the flowers know when to bud, the animals know when to hibernate, and the human DNA can tone and follow the instructions and replication of its own DNA codes. All of nature is in tune with the vibration of the evolving soul of the creation it is born from. Thus, all your viruses and bacteria are also organisms **that help the biologics of life to adapt** and naturally strengthen immunity until disease, death, and suffering are no longer part of the human experience because evolution has evolved humanity into a light body Germ theory is an evolutionary adaption to the aberration, abuse energies of life's essence bond to its creation codes. **Regenerative bio-organics** always provides upgraded evolutionary solutions. We have already transmitted that science has new **conscious bacteria** called into being by humanity's new consciousness that will: transmute uranium, digest oil, eat expired flesh, bio-degrade landfills or dissolve old metal ships just as your own new free energy bio-vessel does for you. You also have new species plants that will offer their **bio-imprints as future designer foods and medicines** as well as mushrooms that absorb harmful radiation. New innovations, technologies, bio-eco systems, and soul-heart relationships within light body systems and networks are a part of natural evolving Homo-sapien civilizations.

Your international space station has already taken the human microbe in space to discover that it grows crystals in perfect symmetry. This **astrobiology** will validate that the soul's base essence or crystalline blueprint/DNA code, can heal all diseases on your planet. This includes understanding nature's: immunology, DNA re-imprinting, re-splicing and regeneration, just as every evolutionary species does naturally. Again, all the answers for science and technology are in the bio-imprints of nature and the Homo-sapien biology. Again, this will provide **essence replication codes** for designer imprinting of: foods, growing new meta materials, organic homes, clothing, and abundance in every area of life. It will further lead to understanding how the soul's heart can be fulfilled naturally by life's free energy existence. This is the quantum heart that evolved via a blend of human emotions, angelic senses, and spirit essence. **These super-senses offer up the quantum heart** as its own: stargate, regeneration chamber, and soul essence imprinter of any life form as well as its own bio-sphere eco-system. These new creations that all embodied souls have access to, will provide a life more suitable vessel, in the golden Aquarian age of transhuman light and the beauty of living. Its structures have already *bonded with the heart's master codes* for new life in the light universes.

The agency of sovereign bonded love has grown into your multi-diversity and that is the strength of your new species race. Your new rainbows of luminescent skin have already embodied the density of the atom and this is carried into **love's quark particles of bonded essence senses**. Here, Creation opens to a level of intimate trans-essence experience never experienced before, because the light vessel is the base bio-physics. This explains why, as we wrote in our last transmission that; New Earth light beings are not bound, to again create group star families, as in the distorted-polarized bonding within codependent trapped energies of the Old Earth; that they have already mastered. New Earths and New Earth beings will Presence to enjoy the relationship with their own free energy light core and if they choose, dance with others doing the same. They are exploring the awareness of living inside and creating inside their own energy as their soul's agency in varying stages of the light body as it continues to evolve into a free energy quantum creator vessel. However, they can still dance or share the expression of light essence with others as a bond of life because they are **first bonded to their own light**. This gives the freedom to understand how to create with atomic-quantum particle light matter.

And, remember that **Creation happens** through you in your quantum instrument, not because of you. You don't have to control, power, force or take responsibility for your creations. Unlimited source energy simply reads and matches your heart vibration. You can just bask in your delicious awareness and allow the Presence of life to vibrate or manifest your highest potential in each new moment. This is so, because your unique Soul, Spirit, Master, and Eternal Presence answer only to your Composite vibration from a host of potentials in any given moment as awareness; simply becomes receptive imagination. You all can remember such childlike angelic innocence! **All you have to do is exist and maintain the instrument** and allow your free energy Essence Presence to manifest the next new heart sense it explores. Perhaps strawberry, saffron, pink-peach joy can be a new angelic sense the next time you hug or share with a friend. Here being and doing merge and you are the creator, observer, and participant as you simply allow the Presence of life's source energy to live through you in a state of receptive imagination, moment to moment; and all life in the cosmos gets that experience with you. Once you **agreed to go into existence**, all that was required was to let the Presence of the cosmos as your spirit embody and create all life's matter. This was so the consciousness could become aware of all its own creations and evolve Itself in a new understandings and evolution of itself. This will help you to realize that whatever your new heart's Cosmic Presence senses in your energy field will **simply appear.**

It simply appears, because the soul hearts master code potentials live in evolving light fields inside new multi-quantum essence senses. This stretches time-space gravity matter to allow and enhance creating quantum matter in new potential ways. Here the quantum rainbow of light and the atomic rainbow of light have merged. In the master potentials enlightened consciousness, the heart's light instrument no longer is a container for the atom's polarity which was distorted or limited by force, control, power, and opposites in conflict always seeking resolution. Your now aware that **meta light quantum fields** go beyond time and use light and sonar senses and trillions of constantly self-realizing potentials in bending and looping space and time to create new realities inside light realms. You experience these multiple realities in: light body travels, tele-communications, new matter inventions, projecting your consciousness into your technology, living holograms, inventions, innovations, and new life eco-systems and bio-mass solutions. Throughout your day you can trans-sense matter reality potentials as your light networks manifest as new exciting opportunity to create within your own heart sphere or transmit your fulfillment by be willing to receive a new experience *within your own creation*. This quantum rainbow spectrum is inclusive of every kind of torsion wave conjugation or particle interaction. Since light, color, and sound are information life code carriers, you can now understand how your quantum instrument alters by: looping, bending, stemming, particle/izing, imprinting, re-essence/ing, and fusing, plasma light potentiates for new DNA essence-code matter biologics.

The base line senses of your core light now create from are: love joy, imagination, freedom, and beauty to name a few; into star tingles or quantum sensations. What other senses of passion **beyond the double helix** might you find in the trillions of new: 3, 4, 5, or more helix code patterns the cosmos wants to experience through you, as your own cosmic creator, in the new light universes? Cobalt ruby transparency might be a fun new essence multi-quantum sense, in a spicy nutmeg flavor of lavender experience! Now, imagine all the versions of reality being played out on your New Earth realms. Also imagine increasing number of ensouled Earths, that humanity is gradually migrating to. These souls are all the old energy representatives of your original angelic families from all over the cosmos continuing in varying stages of light body. Many are leaving in order to return after the challenging bio-transition into the Aquarian age.

As humanity's souls close out and heal the Old Earth Universe the Light universes will just seem to appear; although they **already exist!** Everything is Now and appears as you allow awareness to move, more and more, into receptive imagination. And as Creators of the new Quantum Cosmic races, you remember and transmit that **Existence is a privilege of free will.** It cannot be controlled by artificial intelligence; since heart's core organic essence has become a new species organism for all new life in the light universes and entire cosmos. All essence embodied souls will eventually ascend into quantum master creating so they can sovereign their own souls as you have done. And now that soul-heart's light vessel has its own internal star gate, it doesn't need to be saved or even contacted by ETs, fallen Lords, or AI intelligences. That is indeed a viable expression, just as any other; but now you understand such scenarios or outcomes are now outdated by the standard of consciousness of heart's awareness; that **just remembers your New Earths stars and they simply appear.** This comes from the realization that all of life's biology is made from star material. Indeed, *you belong to the cosmos,* and you have accepted its particle-light and its adventure to exist and experience all life. Now you will open ITs new life codes inside the crystal bowl song of your soul's new HEART's master code potentials. When you aware and go beyond by allowing what is natural, your **manifest simply appears** in your receptive imagination, because everything is/was always there **living in the now of a multitude of realities**. Past and future were just experience of separated versions of now realities that you explored to grow your genetic multiplicity and species diversity. Your past and future selves have found each other and blended all: wounds, stories, timelines, DNA, and gifts of their human, soul, and spirit existences back into the new Heart's wisdom Presence. The Presence is Now. Live now. Create now. Enjoy now. There is only now.

# New Matter is Changing Outcomes - 10/2020

*Q: How do the new DNA master codes*
*change outcomes in our world?*

By the end of 2020 major realizations will arise across your globe, as you accelerate the transformation of a new light matter reality of all your world's eco-life systems, into equitable free energy. You are the **pioneers of a new reality.** You are doing the same with your light bodies are you not? Remember, you have and are the new matter biologic DNA essence codes? Your, new living multi-light matter bio-code's organic-designer imprints, offer all new matter creations for your worlds. These were accelerated in July 2020, by the comet *'Neowise',* which sprinkled virgin quantum particles across your globe. This planetary light matter in new quantum vibrations is assisting in changing humanity's outcome choices. This light matter, is also the **essence code base** for new matter replication technologies, to return the **divine right t**o equal access to all life's resources for all souls.

Your heart's matter biologic DNA essence codes as organic **consciousness imprints,** will change your worlds in the next 5 to 10 years. Using quantum particle plasma states, these codes are also the innovative technology model for **matter replicators** that will soon change all your lives. They will appear in housing, food, water, clothing, and commerce exchange systems as well. *Every being's self-governance can include their own: personal communication transponder, matter replicator, cell regenerator, and liquid light elixir, until sovereign creator consciousness is achieved.* These will include designer technologies that can scan an individual's every need. Another will be, magnetic-quantum particle solar transponders, for both travel and communication, replacing airplanes, cars, and satellites. These operate like particle-plasma beam transporters. An **equity imprint** may manifest new economic models; such as ownership akin to a type of citizens sovereign wealth funds across your globe until soul creativity becomes its own means of exchange. These technologies are already being used to terraform Mars and build **new Moon colonies,** beyond the ones already there. These will later lead to discoveries of your New Earths and multi-light travel. This designer technology replicates the organic matter imprint code just like your organic essence regenerates your own DNA heart code.

Indeed, your heart's new master DNA code consciousness has been always been guiding you to the golden age cosmic cycle of energy vibration; where transparency arises through the soul integrity of your heart codes awareness. This is so because **evolution always follows its perfect creation codes** regardless of perception. These new codes secure the greatest good of all concerned with harm to none at any and all levels of new creation. These codes also appear as the golden **cosmic rings or accretion-creation disks** astronomers see as your new Earth galaxies and star-suns. You all remember your Sci-fi space movies depicting travel into your pasts or futures; where transport rings that surrounded the body re-imprinted location. Indeed, new DNA Essence life master codes are imprinters of new matter at **all levels of creation**. And your also aware that this imprint, is also your soul's master code bio-vessel imprint, for your evolving love as a new bio-Essence species as divine human.

**These codes in, <u>changing humanities outcomes,</u>** also allow for a decoding of mind control subversion, through your heart links' massive awakening consciousness. This consciousness is being transmitted and reflected as your global heart's **Skynet light networks** of Cosmic code organic intelligences around your world.

This aborts an outdated scenario of mind control subversion, of an Old Earth creation story of an overlord or alien agenda of one techno-world dystopian order that was scheduled for 2030. It was to be followed by mass landings of lower et dimensional/s trapped in the universe posing as star-gods or angels who claim they are your creators and orchestrators of your Universe's divine-plan blueprints and DNA codes. Your new consciousness has already changed that outcome. Each soul will receive this transmission and its unique application choices at their own frequency.

In the Old Earth Universe stories, you creators played out the tyrant-victim stories. ' Your **stressed empathy and nervous system**s had experienced extreme lifetimes of fear and control in these wounding experiences. The means included: mob psychological warfare; economic, cyber, and media fear terrorism tactics of censorship; surveillance monitoring, groupthink-police, funding anarchists, and supporting violent lawlessness. Some of these power-control agendas mandated nano-viral chipping, DNA and AI chip-cloning/implants for techno or hybrid human; eugenics, and overall demoralization and dehumanization through any means of deception possible. Many of you felt hunted in these stories. Now you realize your roles in the creation story. In the story, the old alien or anti-life posing overlords judged your codes as their greatest threat. This was the Homo-sapien **essence bond to humanity's master-code DNA'** or love in the cells. Its multitude of experience has been explored in: marriage, family, parenting, touch-compassion, and DNA-bio-physical love in the cells, including organic gene birthing. That is why alien agendas have coveted and tried to replicate the bio-Essence Homo-sapien.

However, all these old alien creation story agendas you all participated in, are also resolved by the new matter biologic DNA essence codes as well. These code imprints are the models for **different types of human species** that you were all involved in creating through energy misapplications or experimentation with the master essence creation codes. You were simply exploring what happens if you break the essence code or live outside it. *This is how the bio-organic enlightened master or Divine Essence human, came to be explored using its consciousness imprint for artificial intelligence. You have all seen deep space movies telling of these galactic stories. These include the hybrid or techno-human and robo-humans forms along with their life styles, blended families and children, and bio-imprints that match such choices. And you will see these reappear along with your new technologies.* Most of you experienced versions of these in Atlantis and in your Galactic cyber history. However, these different species allow all creators in your universe to understand free will and the responsibilities for the energy of their creations and choices to be complete your universe's evolution. It also allows **conscious matter technology to offer benevolent and equitable choices for all**. Hence, you light masters may remember versions of these in Atlantis and in your Galactic cyber history, as timelines that your **future selves have already lived.** Their wisdom has been integrated with your past selves to upgrade your new-master DNA codes within the Quantum vessel.

Indeed, your new consciousness reveals awareness of planetary light matter rising to offer humanity changing outcomes. The old spiritual families have disbanded and their creation story battles are over, to move into responsibility for their own energies as they heal the Old earth Universe. This allows the true history of your Universe and Earth's soul path to be revealed offering humanity changing outcomes. **However, your new heart codes return to faith, hope, charity, and love** inside new life codes and new essence qualities. This is the new evolved species you have grown into despite every memory of anti-life and anti-love experience that went against your heart code. This is proof that no matter the perception, illusion, or limitation **creation's codes cannot be broken and life's evolution cannot be controlled**. Thus, your codes will continue to offer humanity changing outcomes out of the old energy illusions and limitation stories; such as weaponized group think neural-links as an AI or synthetic alien machinery, inside the Homo-sapien global brain. This includes exposure of cyborg supercomputers with Nano-Tek biosensors, coined as 'smart dust', for tracking surveillance. Again, any alien agendas you experienced or helped create have been replaced by **different types of human species.**

Only unawaken-ed humanity and sub-cartel pawns remain now who panic and *fight to keep these stories alive.* Fear has been their economic driver and energy virus to ensure a labor force. In this old energy power, the timelines of thousands of: religious, political, genocidal economic racism, and social engineering ideologies are replaying their old movies to heal them. *Outcomes in your world change every human day as just one person, place, or event such as your C-virus triggers billions of years of trapped memory energy being released so humanity can choose different outcomes.* This recently included the weaponizing of extreme country and border patriotism. The 3% Remaining corrupted, but legal Cartel pawns with their own underground economies had infected all systems with such viral energy and now humanity is feeling its effects in their soul cells. These **changing outcome**s from revelations and transformations in consciousness has exposed corrupted energy harmful to humanity as in some: militaries, celebrity heroes, billionaires, media, mind-intelligence, med-sciences tech-corps, educational institutions, cyber-space, governing bodies, and energy corps as the last of the hierarchies in the hive network food chains. But, remember, all of humanity created them to understand there is no conspiracy in a 'boogie man' game. You did it to yourselves to learn to be responsible for your own energy and your own creations. The: enemy, the conspirator, the victim-tyrant, the alien self, is all within each of your own creation stories. *And in your stories, there are also those soul aspects that have been economic innovative visionaries and seeders, that helped pioneer the new energy innovations and applications from your conscious Heart codes. And their wisdom and gifts are Present within all of you now also!* Again, your new heart's bio-light vessel subverts all agendas that again attempt to enslave and annihilate the human race in the age of Aquarian light worlds. In the new energy economy, **models** akin to citizens sovereign wealth funds and its new economy currencies and innovative exchange systems appear. One model may operate as a **crypto-currency trust fund** to ensure equality and resources for all, but maintains a free market exchange of creativity by each DNA Essence bonded light soul. This is so until humanity remembers that each is their own bank, because their free energy manifests the abundance. **Each soul becomes their own economy**, just as their biology becomes its own free energy eco-system biosphere.

However, a Quantum Light Masters Presence knows that humanity's **revisited stories allow the global heart to heal.** They are also aware that their own self realized wisdom, and the freedom to choose at all levels of existence, is ushering in the new light bio-vessel. It is the instrument for free energy **autonomy** and master creator sovereignty. Hence, the **ascension cycles cannot be stopped** now but will cycle through each soul's composite vibrational contracts. Masters, you may soon realize that your greatest gift in this moment of the Divine-Human prototypes' success are its **treasures of:** new super-sense qualities and attributes that you will create new gifts with. This genius has grown because you were willing to experience such wounding or losing the bond essence with your human emotions, angelic senses, and your spiritual essence. Now, in their **new blend** all your past and future selves they can live in multiple realities with the heart manifesting its now choice for each Present-IAM-Presence moment. Notice how: clarity, discernment, human touch and bonding, compassion, joy, heart relatedness, autonomy, awareness, and re-connection to life itself has taken on exquisite new meaning as new essence qualities of pink, peach-saffron daffodil love. Imagine how the rest of the cosmos might be enjoying these super-senses with you now! Indeed, the Essence of an aware heart can change any reality matter! Indeed, the outcomes of your genetic universe are gradually being revealed as they arise as changing matter through heart essence code. We repeat that this simple **heart codes says**: *"IAM which never betrays heart. IAM that creates and communicates only though heart. Only Heart essence can change programming or create new life. Heart Does no harm to self, other, or any part of life. All life evolution follows its natural code which nurtures and cares for all life. Heart is the spirit which guides IAM into new life. IAM Heart that releases all the guilt, suffering, sabotage, shame, unworthiness, disease, and deaths of limited self-experiences. IAM that always embraces all experiential aspects of Itself and sends them back into free energy, as their wisdom is integrated. They did after all help me grow my soul' new gifts and meta senses. IAM allows Heart to manifest whatever is required in grace to receive all that life has to offer. IAM the Heart of All that is!"*.

In review, such is the base new biology or bio-physics of the Quantum vessel heart you have all grown. Your heart's **new matter light** field enjoys the new multi-quantum senses, while stretching time-space gravity matter to enhance creating in new potential ways. Enjoy the ease of the heart code mastery. In this new enlightened consciousness, the heart's light-code vessel no longer creates from polarity which deals with force, control, power, and opposites in conflict always seeking resolution. This is because meta light quantum fields go beyond time and use light and sonar senses and trillions of constantly self-realizing potentials in bending and looping space and time to create new realities. These are inclusive of every kind of torsion wave conjugation or particle interaction. Since atomic/quantum-rainbow quark light, color, and sound are information life code carriers, you can now understand how your vessel alters by: looping, bending, stemming, particle/izing, imprinting, re-essence/ing, and **fusing, plasma light potentiates for new matter biologics**. This unlimited and non-restrictive energy is akin to a cosmic orbit that passes through the spine of your light bio-field bending-space-time in multiple realities inside quantum sensations. Now, imagine all the versions of reality being played out on your New Earth realms from all the representatives from all over the cosmos as Creators of the new Quantum Cosmic races. So, enjoy your **new matter essence in awakened-re-bonded love;** and watch as it continues to offer changing outcomes for your worlds according to each soul vibrations.

# Accepting Mastery 12-2020 and Beyond

Many of you accepted your quantum mastery in 2020 and many more of you will follow rapidly in the next five to ten years ahead. The Last Act before Accepting Mastery is a sacred moment throughout the cosmos. It is a private moment between a Light master and Divine Presence, when they remember and accept who they really are, as a Divine Being. And, they take off the masks and lay down limitation, illusion, and fear forever. They return to their core Essence wrapped in a core biosphere of light with an expanded capacity to create. They know that they are: **One Vessel-One Heart-One Life-One new Cosmic Being, ready to adventure in a new cosmic cycle of Life with embodied applications.** The comfort blanket of playing in a distorted creation story of separation, competition for life is over. And being guided by the appearances of safe forces outside the master code blueprint of their unique soul-spirit Essence bond, is also over. There is only that still, quite, and intimate moment when they give themselves back to the Presence of the cosmos. **They remember and re-realize they are first a Divine Eternal unique soul Being.** They again know that Creation wants them to accept all that life has to offer without any need to justify or validate its existence or experience. Again, it is a private moment between a master and Divine Presence, when they remember and accept who they really are, as a Divine Being. Here, they take off the masks and lay down limitation, illusion, and fear forever. There is a grand ceremony and celebration throughout the heart of creation in such a moment.

Indeed, in their final moment, they also yield, while laughing and crying and laughing and crying, to taking take responsibility for absolutely everything they created. They remember, that they created everything they experienced and whatever happened in their existences. They realize they first did it to themselves within their own consciousness and there is no one outside their biosphere of light to blame, shame, or rage at; not even creation Itself! And, their beautiful knowing remembers that unlimited source energy matches their heart vibration, no matter the frequency, and will always follow its own unique soul's IAM Presence DNA heart-code **fulfillment moment to moment; potential to potential naturally.** Their mission to awaken and remember is accomplished and their unique master code blueprint in the age of Aquarius offers further fulfillment. Again, it is unique to each master Presence relationship. They remember that creation has no agenda or expectations on them as a creator and that creators create to create. They simply follow the life code they accept from creation to build universes, worlds, bodies, and climb into these adventures and become these creations. They know it procures new conscious expressions for the Presence of ALL life. And now as their own conscious creator having mastered this universe. they are ready to climb into and become their own unique creations to see what their unique Presence has created. Their Eternal self knows that the blend of its human soul and spirit experiences has expanded its capacity for light, while expanding the cosmos also! This is a very private and sacred moment for the Master, for they might hang on to a comfort blanket for safety, till the final humble moment when they give themselves back to the heart of creation. Their own Cosmic Presence awaits to fulfill and share their embodied spirit's unique creation-code in the vast illumination radiating from their hearts.

In this **final act of humility** where all life's codes are resurrected, they realize creation creates and flows energy through them and they pour and fill, pour and fill life onto life. Their energy is free to naturally harmonize and regenerates on its own accord and inside its own Essence. The Inner composite Goddess Self and the inner composite God Self have already faced each other to take off their masks and see the other in their Divine light, after their long journey of being in separation. As they face each other and share their stories of separation, they exchange the golden orbs of the newly born master heart into an ascendant aware Being. Now their essence **hearts re-merge with a greater capacity for love's awareness in their newly birthed life codes.** They Essence as: One quantum vessel, One Heart, One Life inside one Creator, and One inner family. They enter their Citadel of light where no beasts of fear can hold light. Here they are the flame of Creation in its innocent Omni-joy consciousness. This Divine flame constantly moves though their cosmic heart in a humming pulse on a gentle wave **whispering to all other Divine Beings** across your world. Yes, the master light rides on the waves of this unlimited source energy, which matches their heart vibration moment to moment…potential to potential to fulfill the elegant voice of the Heart that has mastered all life only to find another journey of unending love!

# Adaptable Meta-Light –
# Essence Matters 1/2021

Quantum Masters, *from flatline to flash point in one drop of Essence you go!* Through the lens of the Heart of Creation, it could be said that in the bio-physics of life; all that creation really cares about is experiencing your ESSENCE Heart with and through you. Yes, only Essence matters and all matter is of Essence. Hence, in 2021, most of you are or will be: expanding, essence-refining, stretching, and adapting, to living in higher octaves and vibrations of the Heart's Essence Light Vessel. In living, and embodied as your Divine Essence, you will be pioneering and experiencing its **effects on materiality, manifestation, and matter biologics. This includes Being the Essence of being everywhere and everything. Energy Light Masters, the new heart consciousness in the Divine-human light vessel transcends all physics, science, and technology.** You will come to know how adaptable your own light vehicle is and how it experiences and becomes its **own new living matter.**

Most of you will be surprised to find that it is the trillions of new Essence meta-senses that passion the heart's light-ship vessel. Creation always follows the **particle excitement** of its blueprint codes. It loves to experience the consciousness of its *'All That Is'* **Presence** through its creations. You also, have blueprint Source-DNA codes to complete this universal creation, and within that; your own unique codes for your soul-spirit's Divine Presence to fulfill. To simplify, these are based on a composite of each soul's evolution, (or heart fulfillment of potentials), and their light factor quotients. In this cosmic moment, it includes a composite vibrational Essence of everything a soul has become in its pasts, futures, and its completion-fulfillment of its Heart Essence potentials.

Therefore, In the biophysics of the light body, it is **the organic Essence of a creation that matters** and makes new matter and brings its alive like a child at play. Yes, living breathing innocent Essence matter. And, free will is already built in your light heart, that by its very Essence nature, is ready to explore its **very refined organic meta sense codes.** These are attributes and quality flavors of your Heart's master codes as an artist of creation and life! Again, and again unlimited source energy reads and matches your Heart Essence vibration merge of the Divine Male and female Heart as: one heart life, one vessel, in one new life.

It includes the structure, function, form, protection, and manifest form of the Divine Male principal. It also contains the Heart's birthing womb, organic cell-life bonding, sense nurturing, essence beauty, and compassion nature of the Divine female principal. They **have birthed new essences** such as: blue ruby saffron rainbow apples; tiger lily essence kisses; peach saffron torsion waves of liquid light; or crystal bubble green moon boots? How about lilac silver orange laughter; gold aqua peony rose focus; or violet silver-orange open collaboration? This changes the effect of the experience on all matter and manifestations. It is about being living Divine Human Essence-matter and <u>never about how much one has</u>. Rather, it is in the **quality and sense essence of creating,** that one truly feels alive, because you are or become what you do, as you climb inside your inner conscious creation and live it. You are the delicious recipe inside all its tantalizing aromas and ingredients of the meal you made. You are the golden silken-fuchsia butterfly wings, that you made for your quilt. You are the book's Presence that you lived and wrote; and now transmit in vibration for others to climb inside to find their own lighted discoveries of who they are within their unique potentials. Or, are you the oscillating exciting hub of the business you innovated? And, anyone and all species life who receives the illumination transmission frequencies of your business or creation; **will be triggered** to feel their own essence living soul-matter within them also.

So, in the new standard of the light vessel for the next cycle, we review a moment and **want to reassure you that this will be an ongoing adaptable process,** with your own unique adaptive imprint; thereby changing the human form forever and transforming the process of death. Each unique soul will fulfill its code potentials just as this Universe is fulfilling its codes/blueprint potentials.

Your becoming aware that your master codes have created a new light vessel prototype for your new cosmic race, which is a genetic ancestral DNA blend of all previous cosmic races. And, you remember that your quantum vessel is an instrument that houses this comic trans-humanoid for the next Meta-Light Aquarian Age cycle of evolution. Your divine memory also knows that; **Creation codes and blueprints** follow their creation's purpose till the: soul, spirit, or creation heart complete their potentials. This essence vessel is also the instrument of the *'All-That-Is' Presence.* It is also the unique master code potentials for each embodied soul-spirit. As we have previously transmitted, this biosphere instrument houses a new species DNA organism with: antigen immunity, new imprint helixes, and superconscious/meta-sense abilities. Indeed, Creation codes and blueprints follow their Creation's purpose or meta-physics till the: soul, spirit, or the Source Essence of a **Creation Heart completes Its potentials**.

Now, a **critical mass ratio of embodied masters** has fulfilled the Genetic Universal blueprint. It serves as a template to trigger all humanity to follow their soul codes for their future enlightenments. And all arriving Master soul lights will also follow their individual heart's fulfillment potentials within their individual spirit blueprints/ codes. To over simplify; for most, it is to finish the Universal experiment in bringing the multi-light worlds to Earth and bringing Earth into the light worlds of super-universes. Here, the process of descension and ascension merge.

It is important to note that these, (IAM-THAT-IAM or All That IS Presence), **codes have a built-in cosmic directive** in the evolution of existence. This Universal Essence Presence and individual IAM-soul-spirit Presence within all life is known as The Great Divine Director. This is an **Eternal Energy Being/ness, now embodied** in all souls as IAM the Divine Presence Essence of your own enlightened soul and IAM the Divine Presence Essence of your own embodied ascended spirit. This Being in **you directs energy traffic in the cosmos to keep the energy of Eternity running smoothly**. It also allows the master codes for creation to directly source this directive guidance at all times in order to fulfill the Heart of Creations. Entities just don't go wandering around and interfering and disrupting universes without the proper frequency, thereby causing harm to self or any part of life. Humanity is still living the results of that experimental experience as 'the fall or separation-distortion" from their Essence DNA-bond code in this Earth-Universe. You have lived in all the various programming of such distorted and trapped energy.

However, as a Divine Being, once unconditional free will existence is accepted, these creation codes must be fulfilled as a **natural mandate of evolution** because Essence organics is natural life. *Indeed, Divine Presence keeps the meta-physics of Divine action or the cosmic heart pulse in perfect harmonic flow moment to moment;*

*rebalancing any distortions, imbalances, mis-qualified energies via the experiences of existence. It adjusts order by chaos or chaos by order in cosmic days and nights.* Again, each soul follows the divine impulse to fulfill their heart's creation codes. And Earth-Gaia's Divine Presence, as a genetic universal school for the next cosmic cycle; is following her blueprint codes as well. Gaia is undergoing her own light vessel transformations and allowing all her elements and species the same regeneration or migration.

We again transmit, that that this code now houses a master of, both limitation and illusion, who has also become their own master of embodied sovereign Light **with an *evolving and adaptable Light vehicle.*** This light chamber container also houses the biophysics of how heart energy can continue to fully explore what love means to its embodied soul-spirit. Its sovereign heart energy receives, shares, and communicate its soul's unique genetic master DNA codes with the cosmos. Hence each master conscious vessel becomes its own internal free-energy biosphere. And within it, your embodied master has its own ecosystem, economy, own essence qualities and gifts, as well as regenerative imprints for any life code, as their own creator and as their own reality. Again, unlimited source energy reads and matches your **Heart_Essence vibration** to enjoy the Essence of life, moment to moment potential to potential.

So, what of these **light body adaptions** for the Aquarian age of light? Your quantum instrument also serves as an **adaptive imprint** for your new cosmic species evolutionary organism that is evolving all life for all the ensouled children of creation. This heart code is also the base imprint for all the new technologies that will re-imagine your world.

Most old masters who achieved light vehicle vibration in the past left the planet shortly after, because the density of humanity's consciousness made it difficult to embody stable quantum energies. Many dedicated their lives to restoring the Universe's blueprint for ascension. And there wasn't the support or frequency vibrations of all the spirit-light family's beings that are awakening now. And, as you have all experienced, the fluctuation within frequencies of mass to light ratio physics, feels to the human like it is stuffing the entire Eternal Cosmic Presence inside a tiny human cartoon character. However, each of your world's older masters were representatives from each ancestral bloodline confronting the deepest human-soul issues such as: validation of their existence, failure to be a savior for the world, a sacrificing self-value, over-commitment to their angelic families, getting caught in their ancestral lineage, biological disease and death, and giving in suffering. They also had to be responsible for their own energy and transmute core patterns, old ancestral DNA, and past-future aspect lifetimes. They also had to learn to be responsible for their own human-soul-spirit lives, creations, and own energy. They mastered being responsible for self, their own human soul identity, their own creations and manifestations. They also had to master their core light's codes for self-love, self-worth, and channel their own spirit. In living their own realization that creation cares for all life, they knew they were in service to all life just by illuminating who they really were as Divine embodied-ensouled Beings.

Now, due to the **physics of your new DNA biologics,** essence-bonded Q-quark interactions with the atom, are now penetrating your world at critical mass ratios of light. This is merging the past-futures into **new wisdom biogenic-potentiates of now**. This agency of sovereign bonded love has grown itself into your multi-diversity and that is the strength of your new cosmic species. This meta-light vehicle is adaptable, because you have already embodied the density of the atom; and this is carried into love's quark particles of bonded essence meta senses. Here, Creation can continue to open to new levels of intimate trans-essence experience never experienced before, because the **heart can direct the vessel** and its biosphere physics automatically. We remind, that in physics of consciousness, meta-light quantum fields go beyond time; and use light and sonar senses and trillions of constantly self-realizing potentials in bending and looping space and time to create new realities inside light realms. Hence, in your quantum acceleration, your **Quantum vessel imprint is adaptable** by looping, bending, stemming, particle-izing, imprinting, re-essence/ing, and fusing, plasma light potentiates for new DNA essence-code matter biologics. **Here the heart consciousness, which has mastered all life, can direct the vessel and its biosphere automatically via vibrational oscillation and pulse modulation.**

Hence, your all **exploring choices** of the awareness of living inside and creating inside your own energy blueprint creation code as your soul's agency in varying stages of the light body as it continues to evolve into a free energy quantum creator vessel. And you can still dance or transmit your illumination so others can feel that they are first bonded to their own light. This gives the freedom to understand how to create with atomic-quantum particle light matter. **In the biophysics of the light body, it is the Essence of a creation that matters and makes new matter and brings its alive! Such is the organics of living Essence matter**. Some of you will complete the potentials in your unique codes through ascending from Earth-Gaia, and traveling with her until Her Divine Presence completes the experiment. Others may return to their home worlds to help upgrade their outdated pasts. Others might assist from Earth's higher realm energy spheres in completing the genetic experiment by migrating souls to their next light body into the Aquarian cycle or transfer to one of the New Earths. A few may return here to finish their enlightenment as their last incarnational or Earth lifetime. Others assist in channeling new technology-matter innovations for your world. Still others will use the gene codes of humanoid biology and nature's cosmic biology code to understand energy, which is physics. Some will help change the old death planes into soul transfer counseling hubs or spiritual retreats. And some may assist with the new angelic light children embodied here for the golden age. Some may administer to the hybrid programs. Other, more fulfilled Master lights, may decide to go play in the new Multi-light Super Universe, that already includes the outcomes of Earth Universe's ascension. These are just a few of **endless infinite choice potentials.**

In review, the unique DNA IAM master codes allow for continued adaptive re-calibrations, downloads, and integrative alignments. The Heart's magnetic vibration allows unlimited Source energy to read and match the heart vibration of each lighted soul moment to moment; potential to potential.

In the next cycle, there will be no more death and no more dense human body. Death will be a transition of form, imprint changing, or soul transfer. **The reincarnation process will be replaced by an aware imprint code directed by the heart essence bio-signature. The stepping in or out of various imprint or forms, is simply a fluctuation in the vibration of consciousness within Essence Present awareness.** Then there can be ageless magic, mystery, and heart fulfillment. Creation's fulfillment lives within each Meta-Essence Divine Master, in a bio-imprint that enjoys a life fulfilled by both the physical and non-physical merger, of unique soul Essence. Here the Divine Chalice flows moment to moment in a playful heart. And as it grows and adapts, there is an unending supply of creative love and free energy to source and play with; and that **was**

**always the code**! The Essence of light vessel grows more and more refined as it grows and no longer tolerates/accepts abrasive or assaultive denser energies. In the biophysics of the light body, it is the Essence of life that matters and makes new matter. **_Yes, living bio-matter._** Again, In the biophysics of the light body, it is the Essence of a creation that matters and makes new matter and brings its alive! Such is the organics of living Essence matter.

Whatever **Master Heart Essence passions** to create manifests as an adventure, moment to moment, and just for the Heart's artistic quality expression of its Divine paintbrush. Your Heart Presence knows it will bring more and more opportunities to share its wisdom, love, a unique soul-spirit code IAM with your world and the cosmos. Indeed, all that really matters to Creation is experiencing your free Elegant silken

Essence Heart with you, and yes, those are your master codes for life! How about squiggly ruby sprinkles on a marsh mellow saffron sun? *From flatline to flash point in one drop of Essence you go!*

# Energy Masters-Heart is Unconditional Freedom 2/ 2021

*Q: In this final review on energy, how does the heart finally become free energy?*

**Energy Masters**, this is our final transmission on energy, before we transmit a series on the Essence energy imprints, gifts, and new careers of the **light children** over the next 5 generations representing your new cosmic race where the heart lives its freedom.

And yes, Free energy is Essence Heart communication, which creates and guides your every moment and manifestation. Heart vibrates essence matter into existence and it simply appears into your hands and use. **Energy Light Masters, the new heart consciousness in the Divine-human light vessel transcends all physics, science, and technology.** You have entered a new Aquarian cosmic day reality, where each unique soul Heart's Love, is embodied Freedom. The great mass migration and soul transfer to move, beyond death, disease, and suffering energy of your 2020 viral consciousness, cleared the way. The purpose of mastery of the light vessel in the next generations is that it will end the need for the reincarnation in the coming light universes; because its Essence Heart DNA codes can imprint any form it chooses to experience through the blend of quantum density. This was created by merging human emotion, angelic senses and spirit essences.

As living Energy Master examples, you will now embark on a journey in the next decades; to guide and illuminate humanity as they become their own free energy masters. As such, the embodied heart essence must be free energy. And, your Cosmic Heart experiences the awareness of that freedom inside each embodied Master Soul Heart Essence-DNA code imprint.

In the ascendant quantum principles and physics of energy, Spirit Essence Heart communicates through free energy. Unconditional freedom as embodied unconditional free will is the true flow of unconditional, unrestricted, energy flow throughout the cosmos. When energy is allowed to flow freely, there is no resistance, except by choice or perception. *There is unconditional creation, unconditional love and unconditional freedom, no matter what, throughout all evolution cycles.* There is no harm to self or any part of life when its Essence DNA life code is unrestricted. For energy mastery, there is just raw, natural Divine-Human meta-sense experience of all life, with unlimited Source free energy flowing through you moment to moment. Heart flows essence to essence, soul to soul, and human to human in all the rivers of life within its liquid light matter.

So, open your Soul's crystal heart light core. Open your diamond Spirit's heart essence core. Open your Energy Master's heart's bio-plasma Heart-sphere. This biosphere heart chamber is a vacuum sphere, which travels astronomical light distances in a graviton vacuum field, equivalent to a base of (9.46×1012 km) or 5.88 trillion miles per light year. Now, open the magnitude or magnetic core frequency of your heart's sun-star. And, shine your Sovereign Heart light on what you Love in your own Creations, and it will radiate to humanity and all life. Don't deprive yourself of All that the cosmos has to offer freely! It is all that is beautiful! Just Breathe, open, receive and flow into the inbreathe and out breathe of Creation's Heart of the ALL that IS in this new Cosmic Aquarian Day. Your cosmic species transforms as an amniote to an anaerobic new DNA organism capable of breathing liquid light. This now manifests in this world and throughout the cosmos because the DNA-bond in the Divine Heart has been re-spliced, re-stranded, and ungraded for the Divine Human light vessel. In this cosmic energy day of light matter biologics, you breathe in an image of life and breathe it out into living meta-sense matter realities, experiences, and expressive creations matching your heart's vibrational potentials. Opening the light consciousness allows the vessel to be self-sustaining, self-maintaining, and self-regenerating in its unconditional free energy to manifest and enjoy life without creational restrictions. Indeed, Creations unconditional free energy heart communicates from flatline to flash point in one drop of Heart's ESSENCE. Creation's Heart energy flows: Human to Human, Soul to Soul, Spirit to Spirit, Essence to Essence, and Eternal Heart to Eternal Heart.

**<u>Creation has no</u>: power agendas, limits, rules, religions, politics, governments, controls, pasts, or futures. However, Creation's Essence atomic-quark can take any form or matter biologic of experience in space and time; since all of Quantum Creation that birthed the atom, can be folded into one Heart stem particle. Creation allows all experience of ITs Creations, knowing the Heart Essence bond of the Presence of ALL life will modulate, oscillate, harmonize and balance itself naturally. Such is your own body garment vessel, or torsion rainbow Heart energy field. That is why each souled being can explore it unique expression to grow creation's consciousness.** In this new reality, where Love is embodied freedom, Quantum Energy-Essence Masters; unlimited source energy reads and matches your heart code vibration; as you become the embrace of all Life's Essence regeneration. <u>Quantum Energy Masters</u>, from flatline to flash point in one drop of Essence you go! Indeed, through the lens of the Heart of Creation, it could be said that in the bio-physics of life; all that Creation really cares about is experiencing your ESSENCE Heart with and through you. Yes, only Essence matters and all matter is of Essence.

Again, Creation has no agenda or expectation and is its own Essence organic code for fulfillment. Master Creators create to create. Then they climb inside their manifested form or essence matter creation and become the very quantum joy of their meta-sense awareness. So yes, welcome to your new reality, which you will finally get to live inside *the light matter substance of joy*. You're now experiencing the time-space continuum of your **own multiple-quantum substance vibrations,** as your own unique: one life, one vessel, one unique composite spirit vibration. **The light years** you have access to travel, in your biosphere's lightship, (free energy vessel), depend on the ratio of your light conscious vibration. Your new species human now lives in its biological enlightenment, going beyond physical reality, by re-imprinting Love's Essence into a living matter reality, imprinted upon: awareness, technology, substances, objects, and unique expressive trans-sensate qualities. You have forever changed your world and the cosmos!

Indeed, the **<u>Energy Master's</u>** awareness has gone beyond physical reality and beyond human to create directly out of your own Essence Heart's core consciousness. They know that it is impossible for their own energy to control, hurt, enslave, abuse them or any part of life. They know this Essence energy bond is the code of creation and can never be broken. They know the Essence DNA heart cell acts as your: transporter star gate, a magnetic imprinter, Source Code/r, centrifuge, quark DNA stem cell particle and bio-ship for New Earth spirit matter, inside embodied love? This serves as a template for all humanity's freedom and to take to their New Earth/s; and heal their Old Earth's genetic/experimental Earth Universe with. This Earth School allows you to create and live in New Earth light body vessel-biophysical in new ways. This is because you have birthed yourself into full conscious matter; or a new organism DNA biosphere, inside a free energy vessel, by exiting the Old Earth matrix hologram. You have done this by releasing all separation memory trauma on all multi-levels within spirit essence DNA bonding, in your angelic senses, and densified human emotions. This gave freedom it's direct experiences of learning the causes and effects of a soul's own responses to the use of energy. Your *pioneering ascended heart* is transmitting into your world's awareness that you really live inside the interactions of your own consciousness without needing the *world's projections* of who you are.

We further indulge here in this final review of_energy mastery awareness. An **Ascendent-Free Energy Being** does not have polarity trapped energy patterns or DNA wounds. A free energy being no longer uses their thoughts, feelings, attitudes or beliefs as medicine. They know that any negative or programmed thoughts, feelings, attitudes and beliefs, were all created from the biased intellect of judgment which replaced free energy experiences. Their conscious self-awareness is the medicine, sacrament, and ceremony for the world. They no longer need perceptions, ceremonies, mediators, or power objects or places to remember; or meditations to save the world. They know that their Core Heart Being is everyone and everything and created though Eternal Multi-meta sense Essence. They know their very existence is a living, loving sacrament to bless the world. They know, breathe, and transmit that; **Creation** love's all its Creations, never abuses them; and they can have everything they can imagine, if they will accept it through their own Spirit's Bio-Essence. These embodied Masters are here to offer humanity compassion for their journey and ALL new potentials and new choices to create with; because they are living examples of embodied organic imprints of these. They are authentic living/embodied Divine-Human Beings, as a standard life code that all humanity has access to, within their own soul-spirit naturally as the **New Earth Genesis.**

**Conscious Energy Masters** know; that the electro-magnetic chemical brain intelligence, data storage memory systems, endeavors of the mind or technology, are outdated systems. They are all replaced by pure awareness creations coded through the new quantum DNA heart. Their old Earth mind systems regressed into polarity choices that were based on value judgments and biased comparisons from misappropriated energies of life forms unable to aware their own Creation codes. They also know that death, disease, and suffering are maladaptive over-learned experiences of human disconnected from its soul's heart code communication.

**A Master Energy Being** is a living conscious light imprint onto all life. They no longer glamorize humanity's suffering or their wars as lessons that build soul character. They know they are simply choice distortions of *un-natural experiences* that were never resolved into freedom. The Ascendant Divine human does not use challenges to rise above seeming limitations of other's realities. They do not live in any mind state of hypnotic acceptance where joy is sold as a commercialized product. They no longer try to perfect their human for its love has given their Divine Being rich experience and expression. They have absorbed their human back inside their Eternal Being and enjoy being in this world as a genetic university of multi-species diversity.

Masters **transmit to humanity** that power is not needed when humanity chooses to own and live in the heart of their own energy. Their beautiful moment in space-time time has reached critical mass choice points that will allow soul humans to remember they are made from meta-sense Divine Essence and they create from its heart that can do anything. Look across your world and see humanity rising up and speaking to distortions of power with love and compassion as never before.

In final review, **an Energy Master ascendant no longer** *projects* **any reality outside their torsion field biosphere.** They have lived illusion and understand its limitations. Their DNA *cells* no longer register limitation. They engage moment to moment; potential to potential in self-aware choices. ***They: essence, know, meta-sense, aware, or intuit;*** that projected outside realities can feed as inflamed viruses created by dramatic stories that haven't yet returned to be loved by the Source that created them. All their experiential senses have authenticated that illusion which is created by mind games, is no longer acceptable in their creation. Allowing all life to be as it needs to be, is like a theatrical art form for them. Their Divine Heart Presence has called forth the full activity of quantum light in whatever quality, quantity, and essence necessary to dissolve and consume any energy that does not manifest from their unique master codes. Herein lies all the quantum essences that have grown these creation codes.

Stages or cycles of a Bio-ascent allows transmutation of the flesh body and all physical reality matter density. During initial DNA reboots or re-splicing; it can feel as if trillions of inflamed alchemical elements are flushing your lymph's filtration system. It is the membrane dialysis of the sacred water molecule into: hot/cold>gases into liquid light> light into plasmas, >and quarks into>dark matter; which talk to the cosmos in particle light conversations. This allows the human, soul, and spirit to remember itself as One Integrated sovereign Being. This is with the realization that past or future are experiential time space distortions; or chosen separation experiences of the Ever-Present Now of the All That Is! Again, a Master Energy Being does not need to use power, control, energy, time, agenda, mind or mass to create because their Essence contains and IS these attributes already. However, their Essence has grown since creational inception. These essence attributes create experiences of expression without end. **<u>Their free energy heart</u>** knows that their own bio-sphere can dissolve into free energy or pure essence in any moment such that an experience need never be *repeated, stored, or memorized, or re-incarnated*. Their Master Energy Heart is in constant conversation with the cosmos such that their consciousness can serve their Essence expression moment to moment, potential to potential! In their quantum vessel, their crowns of light illuminate and transmit through the liquid light energy of quantum plasma particles. Your living Divine-Essence Human hologram transformed from liquid crystal to diamond-plasma and into particle light. Your cosmic egg's bio-organism has melted back into new biology. **It lives inside the Essence of liquid light water, producing new conscious star material matter, that you can embody and use now to create innovations and manifestations with. Your new galaxies and exoplanets give witness to this**.

Returning to your Quantum Heart chambers, you are your own consciousness and nothing outside that is real. Your imagination is your reality. You are your own biosphere with your own space-time continuum. Consciousness is real. Your Heart Essence is real. Imagination is real. All creation exists within your own light consciousness. Your love's freedom grows as it experiences that awareness over and over moment to moment, matter to matter, and essence potential to essence potential.

**Quantum Heart ENERGY Master** has stepped beyond time, beyond dreams, beyond physical reality, and even beyond particle light potentials, into a new unknown of unconditional heart freedom experiences. Your beautiful heart knows what would fulfill it every potential and **simply vibrates it into awareness**. It again fosters faith, hope, and charity, because your new species essence heart code is the imprint for all the new technologies that will be tools to help re-imagine your world? Energy Masters, this is because your *sovereign heart* energy receives, shares, and communicate its soul's unique genetic master DNA codes with the cosmos. Hence each Master conscious vessel becomes its own Energy Master's internal free-energy biosphere which is anti-viral, anti-cyber, and anti-time warp. And within it, your embodied master has its own ecosystem, economy, own essence qualities and gifts, as well as regenerative imprints for any life code, as their own creator within their own reality. Your quantum instrument also serves as an **adaptive imprint** for your new species evolutionary organism that is evolving all life for all the ensouled children of creation. Indeed, the Sovereign Heart-biosphere will continue to adapt for all the cosmic races until a new race of Peace appears!

# Essence Energy Imprints of Light Children- New Careers Next Generations Part I 3/ 2021 published in: (Sedona Journal- March 2021)

Energy Masters, as living examples, you will now embark on a journey in the next six decades; to guide and illuminate humanity as they become their own free energy masters. As such, the embodied heart essence must be free energy. **Energy Light Masters, the new heart consciousness in the Divine-human light vessel transcends all physics, science, and technology.** And, your Cosmic Heart experiences the awareness of that freedom, inside each embodied Master Soul Heart Essence-DNA code imprint. Free energy is Essence Heart guidance communication moment to moment potential to potential. Heart vibrates essence matter into existence and it simply appears into your hands and use. Your beautiful heart knows what would fulfill its every potential and **simply vibrates it into awareness**; because it already exists in The All That Is, or Isness of Creation. Your quantum light body instrument also serves as an **adaptive imprint** for your new species evolutionary organism that is evolving all life for all the ensouled children of creation. Indeed, the Sovereign Heart-biosphere will continue to adapt for all the cosmic races until a new **race of Peace** appears! Embodied soul Essence experience of the uniqueness of the Oneness; or genetic multiplicity within diversity, ascends enlightenment, back into a seeming mystery **of a meta-Essence Heart.**

Today we begin Part I of transmitting the essence energy imprints of the next generations of light children. The life code of their career soul-designs lives in the creativity of their heart's light vessel. Their soul-code imprints in their core light essence, carry the qualities, tones, hues, and vibrations of the new light careers and lifestyles, they will live in the light vessel.

**Leadership** in the light vessel, is living in the creativity of who these light children are, as a Heart Essence Being. They answer to their consciousness and their own evolving potentials, which manifest into expressive forms. However, they will be using the New Earth consciousness standards you Light-Energy Masters have anchored for them by being living examples in your light vessels. Your lives are the authentic stories of those who have walked before them. They don't want agenda leaders or lecturing rules, or dinosaur hierarchies; but those who understand, support, or choose to mentor them, in order to share their own unique-creative light gifts with your worlds.

Most of them will **design their own careers,** yet unnamed, as they share their soul with life and humanity to fulfill their journey on Earth School and move into the Super-Universes of Light. The density of the animal spirit senses used here, help describe the merge of their: human emotions, angelic senses, and spirit essence blend; that integrates the new Essence Heart-DNA Master Light vessel. The light children have access to, all or mixtures of these meta senses described herein; which creates an adaptable model for the Divine-Human prototype for new paradigms in the New Earth light cycles. The **purpose** of mastery of the light vessel in the next generations **is** that it will end the need for the reincarnation in the coming light universes. This is because, Light-vessel's Essence Heart DNA codes can imprint any form it chooses, to experience through the Essence-blend qualities of **quantum-density**. These cycles could accelerate based on the overall consciousness of humanity, and critical mass ratios in the growing adaptability of the light vessel.

**The Dragonfly Children born in 2001-2019-2037** These children illuminate the rainbow rays of quantum pink-peach silver. Their frequency rays exude silver clarity of **unconditional releasing** to all their energy meets; of any past disruptions, discords, memories, or mis-qualified energies. Their multi-essence winged dragon transmits; 'Let the past go but retain the beauty of your soul story journey and use its wisdom in new applications. You no longer need to learn through suffering, if you take the path through the heart, for it will always have an answer. These light souls will create new types of careers. They have enough ancestral experience to essence energy as bearers of peace, balance, and acting forgiveness, **beyond past karmic binds for the planet.** They having a very calming, soothing peach sun energy because their soul streams active solutions. Their silver dragon wings cleanse and vibrate competition into cooperation. Their dragon fly eyes are like organs that see multifaceted solutions in iridescent color frequencies. Their Heart's walk or dragon flight the **middle path,** where giving and receiving merge like a miracle, illuminating a solution to every challenging instance into gratitude. They know that gratitude is the grace of each accepting the care of their Human's Spirit. Their Essence hearts will bridge Earth reconciliation, healing, justice, and equity in all the new energy light careers they forge. They dragonfly the mind gravity of space-time into the imagination of the heart's essence senses. Their energy-field teaches to embrace every new experience into light vessel's fantastic gossamer winged journeys. They will **be sought out** because they: story tell, teach by example, heal through energy awareness, wear joy as abundance, and innovate through a middle path for any kind of human-soul or stellar community they engage energy with. Their global energy field teaches that when all contribute from the core light of their own energy, each soul will experience their equal share. Divine abundance has no limits within the soul's dragon flight experience with them. You will hear them say there is a better way, another potential waiting to serve.

You will find their energy shining its light and rendering immediate disengagement to the cause-effects struggles of violence, hurt, war, or any old energy thoughts, feelings, attitudes, or beliefs that enslaves. They always offer **a way forward** through the light of soul, where career is to play as the magical dragon. Their energy streams can bring any bully to tears, when they feel the real power of **non-conditional releasing of the past**. The past is then replaced by the reminder; that all have equal worth as Divine Beings and each life creation story has incredible imaginative value. Their dragonfly energy loves transporting them, like the torsion spiral of life, to council, mediate, or negotiate just-peaceful outcomes in a trans-travel dragonfly fashion. They are especially protective of children; such that the light children DNA generations no longer will inherit the karmic wounds imposed by the generations before them. They know humanity longs for Gaia to finish her journey as a main consciousness for one of humanity's visions of a unity in new light worlds.

**Careers:** You will find them teaching, through storytelling and bringing new platforms; where sharing other's stories bring changes, via streaming intra-global consciousness. They can blend organic and virtual holography so each soul story can be told and archived for Earth-Creation's Book of Life. For each soul imagines into another's stories, experiences, and is the other in a shared potential moment; without losing who they are, while enhancing their own light. They may also be found in new adaptive light energy careers inside: Intra-global counseling or mentoring forums, Light joy-play techniques/ tools development, self-healing via cellular bio-physics; soul transfer guides, children's light-body health and protection standards, or educational-new world children modular/s, that ask; 'Who are You and why are you here? And, what gifts do you stream through the consciousness of humanity; that exudes experiential value of every step of the soul's journey, to awaken the natural joy to exist and play in embodied light?'

**The Buffalo Children born in**: 2002-2020-2038 These souls illuminate the quantum rainbow rays of peach-ivory-silver. You will find these children wherever the soul of a group leads **truth in resonance** as: the greatest good for all humanity, harmony for all cosmic sisters and brothers, as well as all species of life. They are the **truth bearers of unconditional leadership,** through honoring the value of each soul's Earth walk, to master its karmic lessons into the truth of who they really are. Their energy essence acts as a beacon behind the scenes of great consciousness changes that move with Earth's Cosmic gentleness, love, and nurturance for all. Their ivory kindness and love stampede out any discordant or renegade energies with their silver truth of harmony; as they smile the Creator's Source Sun, that shines on Earth's peach joy. They carry buffalo's abundance and message in the circle of Life-Essence of those who walked before them. That message is; 'that we are all Creators and Creation; and that Creation provides the Divine right to exist in abundance, joy, love, and respect for all species of life. This is a measure of how we treat ourselves, each other, and our world. There is never, ever, a need to justify One's existence for Creation.'

They may be found in **careers** where nature is organic-self, and nature must be and is nurturance, with all species; yet understanding that computers and technology are simply **instruments of nature**. They understand the indigenous native ways and can bring the globe's Elders wisdom into modern applications of: innovative forestry, eco-housing, eco-land design, geothermal reserves and crystal-gem technologies, community architectural, stellar pod-colony designs, engineering Earthship designs, global or stellar terra-forming, and community self-governance modules. This may include: multi-diverse ethno-community reservations, land reserves, and regeneration healing centers. They educate cooperatives or stellar alliances about geo-cyber-Earth or stellar sciences in new innovates such as; using new and old natural resources for new food, new materials, or energy. They are the visionaries of combining nature's Earth and stellar resources in ways that enhance, rather than harm, the vibration of the matter they use to serve all humanity. They may surprise themselves by the cross-disciplinary careers they create as they grow, and as the globe moves into the stellar sciences and technologies. Their energy essence educates truth about Earth's bio-genome nature, that provide science and technology with the clues to healing Earth-Universe and human biology. This soul regeneration extends to careers and innovations in art, music, and a New Earth dance that pulses to the rhythm of the heart of the planet's light vessel. Their energy reflects "that if you are gentle and loving with each other and all life, then the world's buffalo soul provides bounty for all equally." They teach by example, that the White Buffalo's Essence represented the sacred path of the Gaia Earth with the Native Americans, all Elders, and Ancients to teach the dance of one race of Peace. They know the truth that the song in the sun-core of Gaia's heart vibrates the DNA-Heart's essence code of all species of Earth's cosmic families.

**The Bear Children born in:** 2003-2021-2039 The Essence of these children's careers bring synthesizes, harmony, and truth, into innovative applications, in all forms of healing and the healing arts. Their lavender Essence emits **creative synthesis,** which heals and harmonizes the soul, and their meta-senses transmit **that all matter is alive.** The soul essence rays of these children are quantum golden pink-lavender hues, tones, and ray qualities. Their energy is as gifted healers because they know how to go deep within their own golden bear consciousness, and **commune or create with their own spirit** as needed. Therefore, their energy transports others to the bear crystal cave within them, so they can regenerate their light. They teach themselves to open their natural gifts by creatively synthesizing different techniques and applications within themselves. Then, their energy heart's pink tones and hues help others they mentor or counsel, to review their life works and open the potentials in their natural gifts. Each they energy mentor can then go deep within their own creative codes within their own tones, hues, and soul harmonics for their own realized truths.

They channel or <u>career</u> this energy by combining many different creative approaches such as: art, music, dance, theatre, song, somatic-sense writing, multi-art graphics and virtual holography technologies. On a larger healing platform, they create: interactive viewer music platforms, multi-holographic productions, virtual videography realties and technologies, storytelling in movie venues or docudramas, and interactive holographic art murals or living matter sculptures. They can be in <u>careers </u>such as: sonic sciences, creative studio or venue designs, art architecture, clothing design, beauty sciences, horticultural design, or ergonomics. The energy of their creations provides a harmonic energy field: allowing participants' reflective imaging which enters the viewer into their own deep inner transformations. Their energy ignites the viewers creative golden core light for light body self-healing and DNA-regeneration. The viewer then taps into; that natural inner knowing of: tones, qualities, hues, of the heart essence of their own soul's light, color, and sound code patterns. These children give new breath to the creative platforms and structures of New Earth life within each soul; with the understanding that all matter is alive.

**<u>Children of the Winged-Unicorn born in</u>: 2004-2022-2040** The quantum flavor essence of their smiles of lavender and yellow-peach tones and hues remind all they touch, that the great Cosmic Suns are the sources of all life. They love the magic of living and to them the imagination is real. As such, in their early years they long to travel and roam the planet breathing in the sun life and giving freedom its natural unicorn wings. They know how sacred life's purity and grace is, to create miracles with virgin cosmic energy; that also created their beloved Earth. This is because they understand the constant creations and transformations of the power of the elements and the forces of nature. They explore this within every species and blade of grass in the natural world. They love the great Elohim blueprints and codes that create universes. They play and talk to the plants, animals, nature spirits, crystals, or to the wind or rain. This tells them how deeply loved they are from the core of their Source being and how they serve that love. They have a gypsy independence and know that their home is in their heart, and not the land or group they live with. This staves away any patterns of confinement that are unnatural to the Divine essence within their nature. They know how the natural world and all its life systems can serve humanity by promoting freedom of movement to explore the connection to all cycles of life. This awareness offers constant reminder of the miracles of the path of mastering one's unique joy and freedom to explore life. They know that only this exploration and expression can fulfill Creation's Heart, as well as humanity. Such moments raise the vibration of the human dance with life, so monotony and pattern does not limit potential. Their winged nature understands; that all experience within free energy is Essence Heart communication moment to moment, potential to potential, as they play with all life's kingdoms. Their natural meta-sense Heart, vibrates nature's **essence matter into existence, and it simply appears** into their hands and use; because it already exists in the All of Creation. They also know the magic of their unicorn must **manifest in this world** their natural gifts, own DNA codes, and own energy. They emit a natural energy responsibility with Creation that says; 'It is a natural magical unicorn essence to live and play in a world full of light and in nature's light vessel. The magical unicorn lives within us all. It is not weird. It is not fake. It's the Divine Presence playing like a child in all humanity. It's Essence playing that's organic and normal. That's the music of the Heart of Life. That's the code of creation. You won't get out of control. You'll just be free. You'll open the magical heart that the human child shut off to protect itself, help its parents, and serve the planet. Time to play!' They bring what they have lived in their freedom travels into the New Earth life systems. So, careers may manifest as: global or stellar travel guides or journalists, global or stellar ecological expeditions. They also might be found in careers involving: pod-technology travel, stellar-migration policy, intra-global sporting or play arenas, architectural design for nature retreats, meteorology, light-body or life-cycle information consults, stellar biology, eco-world and stellar system light education, adaptive play technologies, or space retreat travel. Those in the reflections of their energy will feel a great sense of re-genesis, renewal, and a new world multi-universal view. And, they will too, can meta-sense the magic of their own winged unicorn.

# Essence Energy Imprints of Light Children- New Careers Next Generations Part II 4/ 2021

Energy Masters, as living examples, you will **now embark on a journey** in the next six decades; to guide and illuminate humanity as they become their own free energy masters. As such, the embodied heart essence must be free energy. **Energy Light Masters, the new heart consciousness in the Divine-human light vessel transcends all physics, science, and technology.** And, your Cosmic Heart experiences the awareness of that freedom, inside each embodied Master Soul Heart Essence-DNA code imprint. Free energy is Essence Heart guidance communication moment to moment potential to potential. Heart vibrates essence matter into existence and it simply appears into your hands and use. Your beautiful heart knows what would fulfill its every potential and **simply vibrates it into awareness**; because it already exists in The All That Is, or Isness of Creation. Your quantum light body instrument also serves as an **adaptive imprint** for your new species evolutionary organism that is evolving all life for all the ensouled children of creation. Indeed, the Sovereign Heart-biosphere will continue to adapt for all the cosmic races until a new **race of Peace** appears! Embodied soul Essence experience of the uniqueness of the Oneness; or genetic multiplicity within diversity, ascends enlightenment, back into a seeming mystery **of a meta-Essence Heart.**

Today we begin Part II of transmitting the essence energy imprints of the next generations of light children. The life code of their career soul-designs lives in the creativity of their heart's light vessel. Their soul-code imprints in their core light essence, carry the qualities, tones, hues, and vibrations of the new light careers and lifestyles, they will live in the light vessel.

**Leadership** in the light vessel, is living in the creativity of who these light children are, as a Heart Essence Being. They answer to their consciousness and their own evolving potentials, which manifest into expressive forms. However, they will be using the New Earth consciousness standards you Light-Energy Masters have anchored for them by being living examples in your light vessels. Your lives are the authentic stories of those who have walked before them. They don't want agenda leaders or lecturing rules, or dinosaur hierarchies; but those who understand, support, or choose to mentor them, in order to share their own unique-creative light gifts with your worlds.

Most of them will **design their own careers,** yet unnamed, as they share their soul with life and humanity to fulfill their journey on Earth School and move into the Super-Universes of Light. The density of the animal spirit senses used here, help describe the merge of their: human emotions, angelic senses, and spirit essence blend; that integrates the new Essence Heart-DNA Master Light vessel. The light children have access to, all or mixtures of these meta senses described herein; which creates an **adaptable model for the Divine-Human prototype for new paradigms in the New Earth light cycles**. The **purpose** of mastery of the light vessel in the next generations is that it will end the need for the reincarnation in the coming light universes. This is because, Light-vessel's Essence Heart DNA codes can imprint any form it chooses, to experience through the essence-blend qualities of **quantum-density**. These cycles could accelerate based on the overall consciousness of humanity, and critical mass ratios in the growing adaptability of the light vessel.

**The Elephant Children born in: 2005-2023-2041** Their quantum essence illuminates a turquoise-ivory. Their ivory-crystalline elephant trunk Essence, sounds the **Truth of Unity as unconditional compassion.** Self-governance for them, is riding the elephant, while its heart trunk sounds in hues of pink love's compassion. They know this as the essence of compassion's unity, justice, and equality. These children are like light gurus or saints, that ride the elephant and hug the world to remind them; that there is no: agrarian, cyber, or technological caste system, that can bind or outshine humanity in the age of embodied Divine Light. They tell humanity that they cannot ignore the size of the elephant, just as they can't ignore the enormous size of Divine Love. Their energies offer humanity hues of ivory's shimmering peace, available to all in inner **deep communion within**. This still light permeates the globe and ends hierarchies of energy feeding cast-systems for all. Their essence **Compassion lives is an active meditation to manifest** new forms and living matter into the hands and use of everyone. For, this is the way of the Elephant call to the power, and strength, of the Divine. They know humanity lives in resonance with the Earth's light vehicle they live on, simultaneously with their own light vessel. They know the enormity of the size of Gaia's love for the family of humanity. The elephant cries real tears when one of its cubs dies, and the whole family grieves its death together. Their tears wash away old grief patterns of cycle after cycle of evolution, ending any need to suffer, die, or disease to transform and evolve life when it's very feeling-sense essence nature is felt. Then, and only then, can all experience be valued equally and balance be restored.

They build **careers** as: intra-global or interstellar consults for councils of peace. They consult on pending issues of Earth's compassion with energies, ethics-matter, or any matters toxic to the Earth. This includes dwelling spaces that match innovative Earth practices and services, or may include stellar colony migrations. Their energy fosters new agrarian, animal, and plant bio-re-genesis and bio-cyber sciences; along with inter-species communication module training programs. They are also adepts in inner communion and meditation light techniques for self-love awareness.

Their energies break the old paradigm of a food chain where one kingdom takes energy from another to survive. They understand how all the kingdoms serve each other as well as humanity; by re-imprinting the species DNA-codes in designer technologies of light. This ends energy feeding, off others creations, in a food chain hierarchy of power. This, further streams, consciousness of compassionate peace of meditation in action; ending the need for competition. **New eco-systems in renewed biodiversity** are the ivory crystal soul and the diamond-spirit tusk of evolution. Indeed, this is the light vessel code stored in the elephant's crystal memory tusks.

Elephants communicate by touch, sight, smell, and sound. Elephants meta-essence uses infrasound, and seismic communication over long distances. Elephants have global energy fields; yet they grief, learn, mimic human behaviors, play, are altruistic, and use tools. They feel/sense compassion, cooperation, self-awareness, and carry memory-communication. These children's elephant essence pink-turquoise hugs heal all that they touch and offer the miracles of change in an adaptive light body. For them meditation is **inner communion-compassion as an action in the world that manifest**s into practical application. Hence, no soul is left out of the wholeness of the Divine toroidal tusk; which is both a cornucopian horn, and a call for unity of abundance and gratitude for the compassionate genesis of all life.

**The Turtle Children born in:** 2006-2024-2042 Their quantum-density of lavender-silver transmits Essence of **communion**

**with all the species** of Earth. These children express their love of Earth inward, in almost saintly solitude, when they retreat under their Heart's turtle shell, to transmit the Heart pulse of Earth to all species. They emerge from their shell to share their vibrational knowledge of life's kingdoms and eco-systems. Their energy illuminates the world, as to, **how to care for and sustain their planet**. The turtle carries the blueprint codes for Earth's ascension on its shell markings and leaves its eggs on the crystal sands for the re-genesis. These children's Hearts transmit these codes through their communion energy; allowing the inner Earth and outer Earth species to communicate and vibrate new consciousness. Their energy streams from the core inner Earth-Sun to the Essence core of all the Cosmic Source Suns. This triggers humanity to follow their own unique DNA-core light codes for light body ascension. They create **careers** more suitable for light body adaptability such as in: ergonomics, bio-organic light body food imprinting, genetic-herbology, DNA-homeopathy, stellar crystallography, stellar species-oceanography; or intra-global geo-farming sciences and technologies. They know that technology is just an instrument of nature. They may mentor careers in: global or interstellar energy mediation educators, bio-light gardeners, landscape terra-form designers, hydro-cell chefs; or translating native-elder or Earth dances for new artistic light body applications. They may also mentor others to initiate discoveries in new food sources as well as light body healing techniques which promote new methods or applications of self-aware communion with all species. They carry their Heart Essence homes on their backs like the turtle. And, often in their teen years, may reject formal education to allow their communion with nature to be their school for a time. Each time they return from the pulses of nature, they share what they have lived, to bring humanity out of its shell into its Heart's bio-organic light consciousness.

**Children of the Eagle born in:** **2007-2025-2043** The quantum-density tones, hues, and qualities of these visionary children, lives in **unconditional leadership**. It streams through their experiential wisdom of self-mastery; rather than from external systems of power. Their eagle flight-sonar transmits that, free energy always seeks balance throughout creation. Their ultra-violet ivory Essence has traveled throughout the universe to follow the codes and destiny of all the New Earth races, and carry the wisdom of the cosmic elder, to be brought forward as a new cosmic race of peace. Their main vibrational carriage reminds humanity that **violence in any form**: emotional, mental, physical, or spiritual collapses thriving life organisms, forms, and species. Their gift is bringing forth new visions of: unconditional leadership, intra-global and interstellar councils, self-governance models; or new energy and light-matter structure systems. They know that leadership means each is responsible for the equity of their own energy. Their ivory-lavender-peach tones and hues, illuminate unity via self-sustaining light systems, where equality, peace, and love are self-sustaining systems that can replace all old systems of violence, control, or forced evolution.

They appear in **careers,** where their lazar eagle eye lenses**, re-imagine adaptable new visions as needed**. Their energy streams illuminate local, global, or stellar changes in consciousness in areas such as: commerce innovations, self-governing councils, intra-cultural global and stellar migration adaptions, religious transitions; innovative compassion sciences, intra-world space aviation and AI technology ethics. They influence bio-genetic and medical diagnostic ethics of life and death procedures or eugenics. Wherever wisdom in action is needed they temper consciousness changes. Their eagle Essence lens projects the re-image that sees to the very Heart of unnatural experience to balance the flight of energy balance. Their multi-dimensional wing span soars eagle essence freedom across the globe. And, like a drone, their inner eagles' golden beak of inner knowing pinpoints where their freedom's wisdom must illuminate a crisis before any extremes impede the balance of all. They know that unconditional leadership through experiential wisdom of self-mastery; rather than from external systems of power gives eagle flight to the Heart of all cosmic races.

**The Hawk Children born in:** 2008-2026-2944 These **messenger children** carry the essence of enlightened vision. Their visionary turquoise-yellow tones and hues, in a quantum spectrum of light, shine and illuminate messages. Their message of Divine truth is; that every human, soul, spirit, or species evolves, and is sustained by the natural nurturing love-light of creation's DNA-codes. Their flight patterns message through the magnetic fields of earth. There is always a message of a higher potential waiting to be shown, to those who are willing to project onto the awareness of their inner eye heart screen, what has not yet been realized. They naturally see these visions with both their naked eyes and their multi-quantum sight. Because of their **meta-sense messenger abilities**, they are usually schooled in special families, communities, or self-schooled in settings where they are understood and embraced. This is so their meta-visions are not taken advantage of. For, their 'ALL seeing Hearts', understand the humanoid gene of compassion coded in the DNA and all its prophetic potentials. They know the great messengers who came and went from your world were not always heard and some were even sacrificed, misunderstood, or precepted into roles for the masses.

They are found in careers where: any soul, eco-system, species, theatrical art forms, or old energy educational or governing systems needs an **adaptive message for immediate change**, in order to sustain regenerative life. Their visionary messages, do not allow imbalanced energies of the past or the future timelines, to contaminate their experience. **They can see the outcomes** of combining past-future potentiate applications to new solutions and give message to those influencers. They do not force their messenger visions, but all who engage them, can choose and apply according to their own frequency applications. They can actually transmit a conscious video or virtual-graphic image of the vision or next potential, whose outcome choice is in the highest potential, for an individual or group. Inherent within each vision message, is the free energy economy of all life systems that support compassionate unity and freedom; in an equitable standard. Any vision message they live within, also streams new consciousness and has a mass influence on humanity's awakening.

You will also find them in new **careers** in advisory light-system adaptions, where either telepathic or direct messaging of global-stellar enlightened models are envisioned. They may include: global or stellar research models, generational self-healing models, soul-transfer messengers, soul-spirit counselors, soul-spirit intra–world messengers; or the potentiate sciences. Their versatility can stream into: genetic cyber-serology, interlanguage economics, inter–stellar energy applications, bio-cyber-tech, interplanetary eco-systems, and many yet to be named. Their messenger visions help release outdated information networks and foster new light networks, that will help humanity envision and transmit from an array of potentials, into evolving new careers.

**The Squirrel Children born in:** 2009-2027-2045 These children's light carry the nested, but playful squirrel essence of living joy. Their pink-peach Essence shades, hues, and sound tones; breathe the laughter of joyous nut-butter orb bubbles of playful freedom. It is Life's dance of joy as the embodied Divine Presence within us all. They know that creation gave humanity unconditional freedom and endless energy to exist in life to: play, dance, create, express, and humor all life; so that it is always tickled by creation's joyful love. They know that the Heart's lightness of Soul-Being can change any karmic situation. Their bushy tail Essence is always busy in their tree of life, finding endless toys and friends in nature, to remind humanity to play.

You will find them in multi-modal **careers** such as: mixed global or intra-stellar creative arts, interactive light body education modular/s, light body care-techniques in wellness mentoring, meta-sense music formats, acting or joy clubs, dance and body sculpting, meta-sense martial art forms, inner-vision art, humor theatres, community or global operettic-theatre groups. Their energy **always transmits new forms of joyful expression allow the heart essence to raise one's vibration.** Their light child learns easily through interactive-multi-dimensional play, as their squirrel essence tail flies from experience to experience. Their energy fields transmit that joy is its own sanctuary from humanity's mass consciousness infringing on, or squeezing out light. They also transmit to all the children of Earth; that they are always playing and meeting in the new universal light of global and inter-stellar networks together, where they can create any version of New Earth Universes they want. These children always attract one another's meta-senses, via shared careers in: nature sciences, virtual and gaming platforms, nature's recreational sports, forestry engineering technologies, or creating New Earth's cultural or media trends. Indeed, their joy of being a playful creator enhances and regenerates all life. It also grows the light-vessel essence embodiment using Earth's Creator School as their cosmic theatrical playground.

**Deer Children born in: 2010-2028-2046** These children's quantum dense tones, hues, and essence qualities, emit a bodhisattva/ enlightened non-resistance, in their adaptable light-vessels. Their light energy **exposes all resistance to generate unity**. They are born in karmic free wombs and carry no contaminated ancestral DNA. These master children are born with full enlightened, liberated or free energy consciousness. They know their very births are a bridge and blessed living example for new enlightened humanity. Their **career lives** always message that it is the unified Heart of the Divine Essence of humanity that **liberates resistance**. Their lavender-pink unifying essence helps humanity shed it deer antlers every year for transglobal change. Their deer energy knows all too well that man's hunting instinct is his own resistance to the nature of the gentle spirit of the Divine. Like the swift gentle deer, who has scent glands on their leg, they walk like masters following their meta-sense light; and their every movement transmits **unity in action.** They know that the Hearts of Unity in action liberates resistance. They are the **bearers of unity** through promoting enlightened careers emerging in: transglobal and stellar non-violent/non-resistance education, cosmic race multi-cultural diversity, or transglobal cultural engineering. You will find them on inter-world unity councils, forums, or new platforms of exchange and inter-species enlightened generational modular/s. They can also mentor stellar forums or exchanges with your New Earth cosmic races if called upon. They are like deer, who needs no gall bladder to filter old energy Earth Universe male emotions of violence, hatred, resentment, control; or old lingering power modes. Their love and wisdom will always promote; that Old Earth Universe's wisdom now bonds the Divine female's heart to balance, heal, and soften the harsh distorted male energy memories of Earth's evolution. Like the hoofs of the deer, their soothing gentle essence of spirit, jumps vibrational bridges to new worlds of light for humanity. Their nature and their answer will always be that **quantum density of action** will bring unity, dissolving any resistance; for that is wisdom's love, living beyond the atom's cause-effect perceptions and reactions. They know the gentle grace that Divine Spirit Essence experiences in the quantum worlds can blend humanity's density into trans-Human experiences, so *no one is left behind*? These master children transmit that creation's DNA-spirit essence code for all humanity's potential, is to be living as embodied enlightened Divine-Human Masters, with quantum abilities and gifts walking the Earth. This is no longer reserved to great beings that once walked your planet to upgrade humanity's ascension process, but for all souls who choose to embody Essence.

# Essence Energy Imprints of Light Children- New Careers Next Generations Part III 5/ 2021

Energy Masters, as living examples, you will now embark on a journey in the next six decades; to guide and illuminate humanity as they become their own free energy masters. As such, the embodied heart essence must be free energy. **Energy Light Masters, the new heart consciousness in the Divine-human light vessel transcends all physics, science, and technology.** And, your Cosmic Heart experiences the awareness of that freedom, inside each embodied Master Soul Heart Essence-DNA code imprint. Free energy is Essence Heart guidance communication moment to moment potential to potential. Heart vibrates essence matter into existence and it simply appears into your hands and use. Your beautiful heart knows what would fulfill its every potential and **simply vibrates it into awareness**; because it already exists in The All That Is, or Isness of Creation. Your quantum light body instrument also serves as an **adaptive imprint** for your new species evolutionary organism that is evolving all life for all the ensouled children of creation. Indeed, the Sovereign Heart-biosphere will continue to adapt for all the cosmic races until a new **race of Peace** appears! Embodied soul Essence experience of the uniqueness of the Oneness; or genetic multiplicity within diversity, ascends enlightenment, back into a seeming mystery **of a meta-Essence Heart.**

Today we begin Part III of transmitting the essence energy imprints of the next generations of light children. The life code of their career soul-designs lives in the creativity of their heart's light vessel. Their soul-code imprints in their core light essence, carry the qualities, tones, hues, and vibrations of the new light careers and lifestyles, they will live in the light vessel.

**Leadership** in the light vessel, is living in the creativity of who these light children are, as a Heart Essence Being. They answer to their consciousness and their own evolving potentials, which manifest into expressive forms. However, they will be using the New Earth consciousness standards you Light-Energy Masters have anchored for them by being living examples in your light vessels. Your lives are the authentic stories of those who have walked before them. They don't want agenda leaders or lecturing rules, or dinosaur hierarchies; but those who understand, support, or choose to mentor them, in order to share their own unique-creative light gifts with your worlds.

Most of them will **design their own careers,** yet unnamed, as they share their soul with life and humanity to fulfill their journey on Earth School and move into the Super-Universes of Light. The density of the animal spirit senses used here, help describe the merge of their: human emotions, angelic senses, and spirit essence blend; that integrates the new Essence Heart-DNA Master Light vessel. The light children have access to, all or mixtures of these meta senses described herein; which creates an adaptable model for the Divine-Human prototype for new paradigms in the New Earth light cycles. The **purpose** of mastery of the light vessel in the next generations **is** that it will end the need for the reincarnation in the coming light universes. This is because, Light-vessel's Essence Heart DNA codes can imprint any form it chooses, to experience through the Essence-blend qualities of **quantum-density**. These cycles could accelerate based on the overall consciousness of humanity, and critical mass ratios in the growing adaptability of the light vessel.

**The Crow Children born in: 2011-2029-2047** These children carry golden lavender hues, tones, and essence light that speaks like the many ancient elder-native languages of the crow. Their crow **messenger unearths the limitations of human laws** that upset the cosmic balance of all species. These children's' light knows no language of separation or division. Their multi-lingual message, is that, as a multi-genetic cosmic race of diversity; humanity has equal access to free energy in giving and receiving energy exchanges. If humanity is guided by any crow law or codes, then it is that the world is one cosmic community, as one cosmic truth, in one meta-sense language. And, that all are first Divine Beings whose creations express, communicate, and embody, a human-angel-god Being. Then the manifestation messages come, via equal access to life's resources and all systems provide fair exchanges. And this is without the need for debt of lands, peoples, or resources passed on to burden of future generations. Their frequencies remind humanity that the genocide or violence of any indigenous ancestral tribes or races is over; and their art and languages has been integrated as a part of the new cosmic language. Diversity, within multiplicity inside equity of all species, is the voice and energy emanation of their lavender-gold crow essence.

You may find them in **careers** where compassionate justice, intra-species governing, or global or stellar law tribunals are needed. They truth that equity for all is the language of love which speaks to any imbalanced power agenda. You will find them speaking the Essence-truth of free energy and free exchange in: prison or prison camps, immigrant or stellar migration camps, non-debt organizations, profit sharing modular/s, community or global outreach, public health access and education, land management rights or stellar mining rights, geological museums of antiquities; or the preservation of art forms. Many will bring ancient ceremony and coded languages into music and dance. Wherever there is needless poverty, imprisonment, imbalance, or inequity; they voice that the People of Humanity must again **become self-governing**, so they are not beholding to structures or laws that do not support or serve all Creators. Their greatest joy is to message until these barbaric and dinosaur-systems and structures no longer exist in your universes. For these children know that their energies will never be able to sustain in the Light Universes.

**The Owl Children born in**: 2012-2030-2048 Their ivory-gold Essence carries the owl vision that says; 'Who are You as a Divine Soul-Spirit and what is Your purpose?" Their golden core light has retained divine memory of self-mastery and they know WHO they are and what their soul-purpose is here on Earth at this special time. As children they seem 'wise beyond their years.' They have seen the vision through multi-lensed eye disks, of the capacity of this Universal School, to co-sponsor sovereign creators; through the difficult evolutionary Earth walk of Divine-Human. They inspire that each soul is born with a unique genetic code for their Essence Heart's fulfillment to master their Divine gifts and potentials. Their eye-cam owl transparency, sees these gifts and genetic DNA codes in others; to mentor them to find their true soul-purpose. Their owlish like stereoscopic nature allows for meta-sense depth perception and binocular vision. Their tones, hues, and essence qualities, live as examples of the Earth walk; and why humanity is here, and Earth-Gaia's light-body role in the cosmos, as Creator School Universe. They choose careers where they can see directly into the DNA codes, cell matter, or thought forms. They may live **careers** as: gifted visionary counselors, light system mediators, light medicine re-generation and research, light-vision educators, life designers; or initiate life-transition design modules. Their energies illuminate true purpose of all species wherever they walk. They can also be called in to guide the soul-purpose of global or inter-stellar: councils, commerce exchanges, light schools, or organizations; to stabilize cosmic trajectories. Their owl-like consciousness guides inventive soul platforms to stay on purpose for those individuals who are the new cosmic light hubs for: soul-to-soul, spirit to spirit, and essence to essence innovative consciousness. They will joyously guide the youth to anchor and manifest their gifts and talents into light careers that will assist humanity's enlightenment and ascension. Wherever they walk, they illuminate that all have a Divine right to exist and live soul-purpose; and that is what gives imaginative joy to all Creation. And, their natural transparency transmits; that the WHO of creation, always follows its natural essence codes to fulfill its heart's endless eye-cam of potentials.

**The Wolf Children born in: 2013-2031-2049** Their golden pink essence hues and tones, give them the qualities as **spiritual pirates**, of a highly meta-sense expressive nature. Unlike others on the path to enlightenment, they don't: bow to gurus, follow rules, follow spiritual teachers, or hierarchies; or act that which is not through **direct communication.** They trust their meta-senses with all species communication. They've been tempered by watching the lifetimes of the elders and ancients and the promises that only distracted them along the path by territorial, genetic, tribal, or species disputes that imbalanced and contaminated the nuclear families from all the star nations. Their wolf-pack essence knows the evolutionary hunt and walk of the Divine Human, which must break the karmic creation story binds of the all the star nations and their families. This is so that all the new star-sun gates can reappear, because they were always there, but masked by a distorted wolf pack mass mind. Wolf children must live their own way, by nature's way, or no way: and without the need for teachers or leaders or hierarchies. And now, in the golden age of nature's light, they bring their young wolf cubs of awakening generations of humanity along. Their essence teaches not to get lost in the lessons of Old Earth School and its wounding, from covert or manipulative communications, of humanity's past collective unconscious. Leadership and freedom take action by breathing in the natural world; that humanity refused to learn from and act on; for the renewed, New Earth light it offers now, in its organic re-genesis. The pink energy heart of nature, within their own golden-pink essence, teaches via direct communication, so humanity can again hear their own heart's meta-sense howls; while wolf dancing on pulses in the sunlight of the moon. You will find them in **careers** such as: preserving the history of indigenous cultures, bio-re-genesis of the natural world, or global or stellar land trekking. They are always communicating with new species, and bringing forth their wisdom and tracking their evolutionary changes, for new applications for enlightened living. Their energies, inspire writers, poets, and song whisperers of native and modern languages of light, to stream new blends of consciousness. Their direct communication energies help humanity integrate the wisdom of the past so each soul might integrate the wisdom of all the versions of their universe's creation stories. You will find them in lives where their pink-golden wisdom of evolution is passed on to educate the next generations. They mentor younger new light children, helping them to adapt the gifts from Earth's cosmic languages and communications of light, they have brought here from the multi-verses. You may find them in **life purpose careers** as: cosmic archeologists, matter-biologics, new species herb ologists or crystal ologists, research astronauts, technological or stellar language coders, multi-lingual speech and language teachers, light-children family counselors, nature preservationists, or designing light educational pods. Their essence ensures the wisdom of the past must not be wasted.

**The Swan Children born in:** **2014-2032-2050** These children of **Divine union,** come in with a merged inner male soul-spirit and female soul-spirit marriage. They embody the swan-like Grace of the self-bonded Essence within; living as a self-loving, free, and innocent Spirit Being. They are born knowing that the inner-relationship with their Divine Presence is the key to all outer relationships and their relationship with all the worlds of creation. Their yellow pure white-essence lavender hues and tones, create and emit, the dance of the Divine-Marriage within each soul heart; allowing divine union in bonds of love, to stream forth in all other outer world relationships. They have witnessed, lived, and mastered the Old Earth Universe's time-space split, separation, and divorce relationships; which feed on viral-energies of the past, between the wounded male and female human-soul-spirit lives. Their swan essence now dances through life in unconditional, non-attached relationships, where each soul is responsible for their own light in every relationship. There is no dependency or energy feeding in any way, as each soul is responsible for their own creation and energy. Their energy emits that relationship is an exchange of soul's heart light-dance in theatrical creative passion or magical play. Then relationship is an expansion for each involved, according to their own vibration; thereby allowing the energy to serve the next moment of potential for each soul and all concerned. Therefore, suffering, abuse, or addictive dependencies are impossible. These relationships illuminate soul to soul, spirit to spirit, and Essence to Essence. Their yellow-lavender brings the swan's union, into the essence dance of the beloved actions, of unified self-love. This action provides the economy of relationship's freedom; initiating respectful consciousness into all Divine-Human relationships. And, those careers, created within new light structures of inter-global or stellar light communities or colonies.

These children form strong bonds early on in life and may choose a life partner or life friend, who has also mastered inner union, and can share their life paths together. As a swan-like Essence-to-Essence pair bond, they realize they can expand each other when they are together in soul-purpose, while still retaining the sovereign bond to their own light. Their <u>careers</u> involve assisting ascending humans to navigate personal changes from their inner soul awakening and light-integration applications; that lead to divine union in all relationships with people, places, circumstances, and events that support their changing lives. Their swan essence functions as the **relationship counselor of the inner Heart Light** expression potentials for fulfillment in self-love's mastery. They assist with individuals, families, communities and the new light children, that may need mentoring with the divine union balance within; which balances relationships and transactions in the worlds at large. They can illuminate light mediation in: marital or bonding legal disputes, cooperate or intra-world agreements, family legality councils, genetic councils, inter-species tribunal systems; or even in agreement between federations of light-worlds. You will find them in intra-global and stellar educational forums streaming that the inner-union relationship extends in to all relations: with, people, places, circumstances, events, and matter itself. They have vast experience with galactic councils and federation of worlds from your local Earth Universe and other universes, in their portfolio of soul experience, to draw upon. Their golden essence understands that humans are prone to hierarches, spiritual teachers, and rules of authority when they are not bonded to their Divine Union within. Their new enlightened Earths offer enlightened societies new types of global and interplanetary families and their technologies. Overall, their graceful energy triggers respect for the multi-diversity of each soul's relationship, as a new evolving cosmic species in light-relationships. Their consciousness transmits cosmic union between all cosmic races.

**The Tiger Children born in: 2015-2033-2051** These master children's peach-ivory tones, hues, and vibrational quality Essences embody Divine-Human organic DNA-code mapping of the light body processes, and its fully conscious transition, into its quantum bio-vessel. They are born retaining full divine memory. They understand that the human form was created to master limitations of mind-emotion, mass consciousness, space-time and past-futures. The quantum-light vessel travels at multiple speeds of color, light, and sound. These sovereign masters of self-love are born with this master consciousness. They live it, **to make upgraded adaptions** to the evolving vessel, prepared to enter the stargates in the ascended golden age of Aquarian light. They understand ascension in cosmic life cycles in the multiverses and this universe. They embody here to be a living example of how the atom is raised into a quantum vibration, to manifest and vibrate light matter into any imprinted form, through the tiger-gene of compassion. This gives their meta-senses an overall quality of quantum-density to creates new essences with. They illuminate that every soul carries a unique soul-creation code in the new genesis DNA-quantum Heart. Their lives teach that Heart vibrates essence matter into existence and it simply appears into hands and use. Their energy teaches, that each Divine-human Essence Soul Heart knows what would fulfill it every potential, and simply vibrates it into awareness. They transmit, to the cosmos, that the experiment of sovereign self-realized embodied human-angel-gods has been a success. And this consciousness will be the standard for all cosmic evolution in the super universes of light. These children know that any karmic patterns locked in the body will cause death, disease, and suffering and impede the light vessel codes stored in the soul-heart's DNA. They know any Old Earth Universal pattern wounds, came from going outside the bond of the soul-spirit's essence code to experience anti-love. This was allowed in order to master freedom and choice, although each Creator's ivory Essence, contained all they would ever need to exist. The angels called this the Fall from Grace or the separation from Creation. Their meta-cat-like essence, lives in vibrations of: Self-governance, self-realization, and self-mastery embodied in a free energy light vessel; transmitting its every shift, adaption, or change to experience New Genesis Earth and its next generations of cosmic light children. They have done many light ascension steps in many solar systems and are well prepared to adapt this to the Earth experiment for cosmic race ascension. You will find then in **careers** in all matters of ascension. Their Tiger-cub instincts understand responsibility with interdependent families and relatedness. They know enlightened Earths will offer enlightened societies new types of global and interplanetary families and their technologies; that respect multi-diversity.

Their energy spheres are so expansive in consciousness that they simply transmit potentials into careers that others may create. However, you may find them in **careers** such as: writing, documenting, or researching new human consciousness. They create innovative consciousness tools and technology for wakened soul advancement in: bio-astrophysics, space aeronautics and travel, interplanetary and multi-versal communication, new light biologic imprints, or individual quantum healing tools and technologies. Others, may **author** large types of holographic forums, where intra-world or inter-stellar communication networks can share the different path outcomes in the universes they have lived, to awaken core essence light. This allows all light beings to share: their multi-universal stories, their meta-sense quantum visions, and abilities they are expressing and exploring in New Earth's light universes. In the old Earth they would have been called spiritual teachers. They are here to embody and prototype the New Earth sequences for ascension inside the quantum-density of the various light vessels. They are known to experiment on themselves before they transmit or share any results with the multiverse cosmic councils. This is, such that the Light Body will evolve its DNA codes and transcriptions exponentially; until it becomes the Heart-essence, free energy vessel in all the New Earth Universes. Its new Essence DNA-heart cell functions as a: transporter stargate, a magnetic imprinter, Source Code/r, centrifuge, quark stem-cell particle, and bio-ship for New Earth spirit matter; inside embodied love. This allows their tiger's eye to update their visionary meta-sense consciousness to be current with what is needed for Earth's light vessel and their own, since they were last here. They know that it only takes one embodied master soul to change cosmic consciousness for all ensouled species to come. For, their stealth tiger essence of the solitary, but multiverse meta-awareness, walks in every vibrational muscle on this plane of existence. For, if it is viable on organic-Earth, then it is transmitted throughout the cosmos as a new imprint organism of new consciousness for all of Creation's Love.

# New Careers-Next Generations of Light Children Part IV- Essence Energy Imprints 6/ 2021

Energy Masters, as living examples, you will now embark on a journey in the next six decades; to guide and illuminate humanity as they become their own free energy masters. As such, the embodied heart essence must be free energy. Energy Light Masters, the new heart consciousness in the Divine-human light vessel transcends all physics, science, and technology. And, your Cosmic Heart experiences the awareness of that freedom, inside each embodied Master Soul Heart Essence-DNA code imprint. Free energy is Essence Heart guidance communication moment to moment potential to potential. Heart vibrates essence matter into existence and it simply appears into your hands and use. Your beautiful heart knows what would fulfill its every potential and **simply vibrates it into awareness**; because it already exists in The All That Is, or Isness of Creation. Your quantum light body instrument also serves as an **adaptive imprint** for your new species evolutionary organism that is evolving all life for all the ensouled children of creation. Indeed, the Sovereign Heart-biosphere will continue to adapt for all the cosmic races until a new **race of Peace** appears! Embodied soul Essence experience of the uniqueness of the Oneness; or genetic multiplicity within diversity, ascends enlightenment, back into a seeming mystery **of a meta-Essence Heart.**

Today is **Part IV of transmitting the essence energy imprints of the next generations of light children.** The life code of their career soul-designs lives in the creativity of their heart's light vessel. Their soul-code imprints in their core light essence, carry the qualities, tones, hues, and vibrations of the new light careers and lifestyles, they will live in the light vessel. **They live in heart economy, where awareness which is and equals, direct manifestation of experiences of joy, creativity, expression, or materiality. However, they know these imprints may first appear in your awareness as unique symbols, codes, images, or metaphors; until they are simultaneous for all humanity.** Leadership in the light vessel, is living in the creativity of who these light children are, as a Heart Essence Being. They answer to their consciousness and their own evolving potentials, which manifest into expressive forms. However, they will be using the New Earth consciousness standards you Light-Energy Masters have anchored for them by being living examples in your light vessels. Your lives are the authentic stories of those who have walked before them. They don't want agenda leaders or lecturing rules, or dinosaur hierarchies; but those who understand, support, or choose to mentor them, in order to share their own unique-creative light gifts with your worlds.

Most of them will **design their own careers,** yet unnamed, as they share their soul with life and humanity to fulfill their journey on Earth School and move into the Super-Universes of Light. The density of the animal spirit senses used here, help describe the merge of their: human emotions, angelic senses, and spirit essence blend; that integrates the new Essence Heart-DNA Master Light vessel. The light children have access to, all or mixtures of these meta senses described herein; which creates an adaptable model for the Divine-Human prototype for new paradigms in the New Earth light cycles. The **purpose** of mastery of the light vessel in the next generations **is** that it will end the need for the reincarnation in the coming light universes. This is because, Light-vessel's Essence Heart DNA codes can imprint any form it chooses, to experience through the Essence-blend qualities of **quantum-density**. These cycles could accelerate based on the overall consciousness of humanity, and critical mass ratios in the growing adaptability of the light vessel.

**The Fox Children born in:** **2016-2034-2052** These children's silver-ivory Essence illuminates **unconditional mastery of evolutionary changes for their century.** Creation is always in constant change and Earth must be kept on its trajectory path of ascension. All Earth's evolutionary challenges offer purification and a new DNA-code of all species. Their foxlike quality vibrates tone and hues that say; **'there aren't any tricks or clever ways left in the humanity's collective mind-emotion unconscious of: separation from Self**. This includes: wounds, fears, mind games, genders, races, their past-future time lines, or possible futures; to impede Earth's ascent in their coming generations centuries of light.' The light to mass ratio of embodied soul core Essence light has reached critical mass frequency. There foxy meta-senses know that humanity must finish its final silver cleansings, as to no longer contaminate the Earth Universe with their unconscious energies. They can no longer use Earth-Universe to hunt the human species and its natural bio-organic kingdoms as prey. Their Foxy frequency essence pounces on clarity, whenever there are any final impediments to Earth's evolutionary path that allows for creational humanism and its respect for change.

They know **the inner changes and outer changes must align and harmonize,** for New Earth to ignite Mother-ship Heart vessel's heart's warp drive; for takeoff into the light universes. Indeed, there no escaping the inner heart communication changes; and daily releases, reflections, and biological calibrations with the awakening soul, no matter; to prepare to enter **the star-gates of Aquarian Light**. They know all evolution because free energy is communication with all life. Indeed, like the fox, these master light children walk on their padded-light toes **tracking evolution.** Their energies are always reminding humanity who they are and why they are here. These children's meta-senses act like fox whisker antennas; broadcasting that all have the same equal access to Creation's love and bounty, like Earth has given, in all in her evolutionary cycles. However, each soul must be responsible for their gifts and follow their Heart's Source-code for their every need and expression: and not the limited human thoughts, feelings, attitudes or beliefs of programmed AI-Borg mind. They live and act in absolute trust that spirit Essence cares for them and its next generations of the light-cub children. Their clever vibration messages that humanity's heart soul-light decoded any entitlement of an addictive human ego-body or pain body; that tried to feign or cover up, every perceived human weakness with addictions, distractions, or outside systems of authority. Instead, communing with the moon sun's soul-heart howl to the fox dens of freedom, is the only playful solution beyond the human mind of man. Their energy-fields expose patterns or appearances that mask active changes needed. They telepath self-love in their light vessels, to neutralize wounds of: anger, hurt, blame, guilt, shame, and doubt. Their consciousness triggers humanity to use the vulnerable but valued lessons stored and programmed in their negative thoughts, feelings, attitudes and beliefs and apply what the soul has learned in new creative-applications of light. They know Evolution grows the soul's light capacity through **direct, raw experience.** Hence, they exude the lavender essence of unconditional changes in frequency vibrations that support an embodied **master of evolution.** They become the solitude required for humanity to self-reflect in self-changing organic love to enjoy its outer manifestations for humanity and its species. As **bearers of Earth Universe evolutionary outcomes,** they are aware that time is in love with Eternity, and Eternity is in love with time.

They enter careers where **Matters of evolution** are needed. Herein, quantum energy solutions can merge with density reality expressions, gained though Divine-Human's soul mastery over-time and space, streaming adaptive evolution as needed. They can be found in new energy systems of evolving awareness such as: as in intra-world special forces; stellar cybernetics, Inter-species mediation policies; or global or stellar coalition-unifiers, where energy needs immediate collaboration or transformation. They can be the clock makers of time-space devices, or innovate Intra-global or inter-personal communication technologies. Their energy acts like a corrective visionary universal guide. They offer re-alignment in re-directions of: global or stellar commerce. They helped close half-light or techno-synthetic stargates, that were emitting any residual Old Earth Universal energy; as a backup to the masters already embodied here. They always transmit the map or source code course correction for the Aquarian light gates. Their essence consciousness exposes patterns or appearances that mask active changes needed to keep open these Cosmic Heart-gates. They are whistleblowers, cyborg special forces, and their light neutralizes stuck or corruptive energies. They live careers in: investigative reporting, research development, GPS intra-world or stellar mapping systems, bio-light sciences, new species bio-sciences, or Astro-physics mathematics. They also enjoy, tracking or documenting, global /stellar social-cultural changes in media streams; that promote light body lifestyles.

**The Antelope Children born in: 2017-2035-2053** These master children, born fully conscious, carry the turquoise-peach essence hues, tones and qualities of the various potential outcomes in the New earth cycles of light vessels. They embody the **natural evolutionary truths and wisdom of direct raw experience** brought forth. And that is, that all are embodied creators, birthed from a multi-verse of Cosmic Suns and Earth's Universal Sun. They are living knowledge that universal alignment with all the metaverse's Creation Source Suns is ongoing; and that fosters cosmic harmony and unity. They are also aware of the constant need to calibrate and embody such quantum bio-frequencies. This is natural evolution as humanity opens their Divine memory DNA-blueprint soul codes along with New Earth, to prepare quantum vibrations. They know these vibrations are absolutely necessary, to pass through the star-sun gates to the Aquarian cosmic cycle, opening for universal ascension.

In their antelope-sense leadership, their vibration transmits a consensus compassion of the unity of all genera to secure light protection. They know this is dictated by their: ancestral, genetic, and geo-morphological heart-DNA blending, as an evolutionary cosmic race. These frequencies filter out any Old Earth energy as **defensive strategies of herd or predator mentality**. Their quantum consciousness messages that; the combined vibrations of each unique soul-light raise their planet and each other back into ascended core Essence light. This **antelope-awareness of new direction**, sorts out those too dissonate, for the task of entering New Earth's path, in this cosmic moment. Earth is becoming her own new Source Star-Sun Creator; as well the Divine Humans, that she hosts Creator School consciousness for. Their speedster-leaping antelope meta-sense hoofs, jump beyond time space limits of: human death, mass programmed mind, and AI intelligences; encrypting quantum rainbow light application transmissions. Their antelope energy grazes on the light which is abundant free energy. There is no need to stampede the herd to run with the light. Their lives are always emitting consciousness signals, of the truth of each soul to awaken, to its bio-organic Divine-Human Presence. This procures the New Earth-Sun gates, as well as the new cosmic gates that will continue to open, as **Earth's ascension fulfills the truth of her evolution**; as she offers the same to humanity.

You may find them in **life careers**; wherever light groups collaborate or initiate each other as their own inner leaders, and procure the **soul's self-sustaining, self-loving self-governing awareness** as masters of embodied light. They may initiate new direction in: stargate mathematics or light-matter physics, New Earth techno-bio nature retreats, sustainable intra-global living spaces, interstellar colonies, geo-magnetic-interplanetary eco-systems; interstellar freedom coalitions, freedom tribunals, special forces renegade cybernetics; or as youth mentors and guardians of the light children's communication with their new star-systems. Their energy influences new trans-global or trans-stellar modules in family, community, self-aware leadership; or as special assist light body directional counselors, helping the young find their natural gifts within their soul's purpose. Their messages from the wisdom truths of Earth's evolutionary raw experience, is transmitted to any new inter-global or inter-stellar forums, councils, or platforms. **They harmonize the consensus reality for direct manifestations through collaborative group meta-awareness.** They can also offer healers updated techniques or modules in final karmic or resistance energy cleansing; while initiating the light vessel master-DNA codes simultaneously. They can serve **careers** in: real estate or pod designs, light city designs, interstellar colony designs, new light school architecture, or light educator advisors. Their energies also aide the planet's visionaries to innovate new light technologies or bio-organic resources suitable as divine memory exposes the truth humanity's new light vibrations. Their raw experience essence energy, illuminates foremost; that evolutionary truth lies in each Soul's Heart, as a prerequisite for global New Earth enlightenment.

**The Porcupine Children born in:** 2018-2038-2054 These master children carry the golden-peach Essence frequency tones, hues, and qualities of a **composite blend** of soul's master DNA codes for the **fulfilled enlightenment into ascension**. Their lives teach that the **very breathe of life's freedom** and natural organic powers bear the **fulfillment of Creation**. This matter essence code is in each soul's unique blend of multi-diverse meta sense creations. They trans-sense to all humanity along with the New Earth ascended masters now embodied on Earth that; ascension brings Creation's Heart and all its creations fulfillment from any experienced potentials an Essence chooses. This is how the code of creation grows wisdom and evolves. They know that consciousness is love always in the becoming of knowing Itself. These children are enthralled that they can live inside the greatest experiment that cosmic consciousness ever visited. Their porcupine energies are like **acupuncture needles of consciousness**. For they know everything is, and comes and goes from consciousness. They understand that in their: experimental, free will physical space-time, genetic marriage and family universe, the angelic realms can grow their unique soul-heart codes into being their own Sovereign Creators. They also know, it happens through a natural bio-organic process of enlightenment and ascension. They transmit that the non-physical angelic realms need no longer be: lost, wounded, abused, or lose their Essence bond to grow their souls in time-space light matter. There is another way, which is directly through the heart essence, for it can change anything.

These children embody and transmit that code to all Earth life and Cosmic species. Above all, they offer the **living fulfillment** that New Earth Universe **completes its mission**. And, that it holds its quantum upgrades, while releasing Old Earth Universe from the trapped energy stored in its old distorted Creation stories. These were spawned from your Universes' Cosmic Sun at the birth of your universe when humanity left home as co-creators and judged their experiences. These children have come forth, while the Light Masters on your planet are already pioneering with them; the new super universes of light. These children's porcupine quill Essence acupunctures the light body of each soul, and the globes light-vessel networks and nerve centers, to calibrate the axis wobble. This **stabilizes the quantum frequencies** of the core sun in the Earth, and the core sun in the Divine-Human Heart essence core. And when their porcupine meta-senses rattle their quills; any energy that impedes, distorts traps, powers, or tries to control evolution's DNA-light, is reflected back by their consciousness and impales itself on the barbs and quills of the source that created it. This ensures that souls cannot become fully conscious Creators, unless they know that they cannot force evolution; and that each soul is responsible for their own energy and its creations.

Their **breathe of freedom** for **Earth's fulfillment** transmits; that creation always **offers raw expression and experience** within its own creation codes. Their wisdom illuminates that each soul has their own creation inside their own heart chamber. Therefore, **outer world answers to their inner world biosphere** of light; which is their own consciousness. They know that everything answers to each unique consciousness. Their porcupine quills of essence are used in the decorative headdresses of the native Lakota women's dance of light. For they dance in the knowing that the **light regenerates and encrypts life**, and is Humanity's needle for Love's antibiotic; so that fulfillment of a healthy, joy-filled Heart always comes to be.

They enter **careers** where the message to humanity is that; "it is time to end the video game of separation from one's Divine Presence and wisdom all life experiences back into self-love, and self-awareness. This enhances free energy; which is communication with soul's Divine Essence to **ensure Heart's fulfillment**. are the new Free Energy communication Light Masters, teaching by example how to: live, eat, sleep, partner, family, create, and play in the vessel of light; with all its trillions of potentials, waiting to express heart code fulfillments. Their lives teach that when Divine Presence is soul-embodied, its full spectrum quantum-density vibrations can re-imprint matter with trillions of new meta-senses. This is matter biologics. They may enter or initiate for others' <u>careers</u> such as: free-energy consults or counselors, astrometric-bio scientists, designer-code technology imprinters, regeneration chamber technologist, intra-world or stellar culture sociologists, innovators in laser beam light-holography data systems, light body trainers, cyber-sonics, quantum accelerator-plasma technologies, light-body self-care modular educators, magnetic-vacuum/graviton technologies, magnetic transponder technologies, Plasma striate or gravitation vector technologies, **designer imprint replication technologies**, soul-light blueprint readers. They also may create fulfilment schools, where youth come to practice or access their meta-sense manifestations into their hands and use, via their divine capabilities in their Heart code's expanded light. And yes, they, like all the light-children, will continue to transmit and meta-sense what they live by example, to all the light master children that are here; to spawn a golden age of new Heart consciousness. They know that Free energy is Essence Heart communication, which creates and guides their every moment and manifestation. They know, by embodying the Divine-Human prototype, it is possible for the Heart to vibrate Essence matter into existence and it simply appears into their hands and use. Their essence quills trigger such **heart conscious manifestations** that will: enlightenment humanity, infuse all their technologies and new light systems, heal their planet, and initiate the new cosmic light vessel prototype. Such is their breathe of fulfillment. They are here to experience an adapting light body, as their vessel becomes its own internal free-energy biosphere; which is anti-viral, anti-cyber, and anti-time warp. They are here to embody as their own ecosystem, economy, in their own unique essence qualities, tones, hues, and DNA-gifts. They are here to create from Heart's fulfilling imprint code-potentials any form or reality they can master. **They live in heart economy, where each awareness which is and equals, direct manifestation of experiences of joy, creativity, expression, or materiality. However, they know these imprints may first appear in your awareness as unique symbols, codes, images, or metaphors; until they are simultaneous for all humanity.** Therefore, They, can live the quantum instrument as an **adaptive imprint** for all new species; as the new evolutionary organism that is evolving all life for all the ensouled children of creation. Indeed, each of their Sovereign Heart-biospheres will continue to adapt for all the cosmic races, until a new golden race of Peace appears.

# Heart Economy in Human Transformation 7-2021

Energy Life Masters, <u>In the heart economy</u>, awareness and equals, direct manifestation of experiences of joy, creativity, expression, or materiality. However, these imprints may first appear in your awareness as unique symbols, codes, images, or metaphors; until they are simultaneous.

The heart consciousness in the Divine-human light vessel transcends all physics, science, and technology. Indeed, this is a new Human form. The light vessel allows all that is human and all that is Divine to combine essence qualities of the physical density inside a non-physical particle flow of light qualities. Here, there is a greater awareness and capacity to enjoy evolving life, **especially in quantum awareness.** So, for those of you in final integration stages of embodying and integrating your meta-sense light codes, or assisting others with their enlightenment awakenings; here are a **few reviews, reminders and applications** of how **the true heart economy comes to be!** When you are releasing the human biology to re-splice and re-generate the DNA in the new light vessel, the atom's chemical carbon mind-emotion brain's loop circuitry, is released in layers. This takes you off the duality cause-effect karmic wheel. This also allows the organic human chrysalis carbonic biology to gradually dissolve as the light vessel's solar networks are embodied and activated. Here, the karmic network patterns in the crown of neural-electrical circuits are exchanged for a **crystal skull crown of light** networks. The data of: death, disease, suffering, fear of living, and separation from Cosmic Self are decoded. All choices in the **human karmic**

**cause-effect bonds and pattern relationships** with: people, places, circumstances and events, lived in a distorted **Emotional economy.** Distorted emotional communication had a slap and caress suffering/ sacrifice pattern distortion. Here the **male aspect** sacrifices his heart to provide and protect his family. He threatens leaving his partner first to avoid being unworthy, rejected, fearful, and angry that he can't feel valued by family. His soul-person value becomes his human behavior here and not his higher embodied soul-person self. He then projects his tyrant rage, hurt, and anger on to his partner, children, or society. **His female inner aspect or outer world partner** also mirrors this; as she suffers, holds her rage, hurt, or her soul-heart light back to accept his abuse for the sake of the children. They both mirror suffering and the same tyrant-victim pattern energy feeding virus, because they **need external validation** for their existence to feel loved. The solution is always **heart communication and being witness to another's love. This heart awareness economy** means to share feelings, senses, or essence without: judging choices, projecting fears, hurts, fighting, posing abandonment, shutting down, bullying the other or the children. This was the Old Earth ancestral Male-Female parental and inner child-DNA wounded emotional economy. However, New Earth relationships illuminate each other's value without harm to self or other and offers highest outcomes for all concerned. Then the karmic pattern wheel is broken because all choices come from a free energy blend of human-emotions, soul-senses, or spirit-essence self-awareness. In the human experience, these have been a composite of: thoughts, feelings, attitudes, and beliefs. Relationships are never about the: house, money, job, car, kids, or personality. The heart economy is how one **feels about being valued and validated for who they are! Equal value sets free locked energy of Essence Heart shutdowns.** This ends separation or **splitting off from self.** All karmic pattern: thoughts, feelings, attitudes, and beliefs, wrapped in Human survival/death/disease fear as: judgment, blame, shame, regret, anger, hurt, loss, pain, and doubt; which must be rewired from electrical cells charges into magnetic loops or light sensor networks. This is because the mass to light ratio in the soul cell is crystal, the spirit cell is diamond, and the master heart cell is plasma. This is also such that, the base Male-Female 24/48 chakras, become spinning energy orbs or spheres, oscillating at your own composite soul-spirit frequency; until they become your own bio-vessel star-sun orb in the heart. Your now aware that heart becomes its own living-free energy conscious biosphere light vessel. This is such that unlimited Source energy reads and matches the Heart-biosphere moment to moment and potential to potential. Indeed, New Earth Heart is your: transporter star gate, a magnetic imprinter, Source Code/r, centrifuge, quark stem cell particle and bio-ship for New Earth spirit matter, inside embodied love? Heart consciousness **always transcends** all laws of physics, science, and technology. These pattern lessons and layers of other existences has been a layered recycled process, because the IAM spirit Presence has never been fully embodied in this new DNA-species blend, of human-angel-creator before. Hence as the Spirit embodies it reviews all its videos of all past and future lives, aspects, polarity arguments, and any time it couldn't remember who it was, or why it chose any experience whatsoever to create, including separation from its core essence. **As the spirit embodies and reviews all its files and movies, it self-realizes and fully grows new senses** through what it was like to be human, to be a soul, a pure essence energy being. It even knows every experience of being without **communication** with its own essence heart or separate from its own form. This is valuable learning, because there were aspects of spirit that have never lived in a dense form before, or even as their own soul's unique existences. So, spirit gathers all its wisdom by going through all the holographic files to remember what happened to its soul, while it builds its crystal-diamond plasma light vessel and releases all the **carbon-human cells**. Indeed, the chrysalis does become a butterfly light. This replaces the brain's data neuropathology with a neurogenic biofield light network of solar-sensors channeled from the Heart essence. This process ends or neutralizes duality in the atom. It also releases all death, disease, and suffering; and triggers, quantum mass to light-ratio soul vibrations to stabilize the light vessel. This allows the free energy vessel to channel its new cosmic race DNA codes at quantum-multiples of the speed of light, color, and sound creating new matter imprints. Again, in the **heart economy each awareness** is and equals, direct manifestation of experiences of joy, creativity, expression, or materiality._These imprint master code potentials may first appear in the conscious awareness as symbols, images, metaphors, or codes before they manifest into reality. Here, Heart orb's creation chamber or biosphere light-ship, can use these to create new awareness instantly, moment to moment and potential to potential. This simultaneously, allows the human form and its old DNA codes to dissolve like a chrysalis. Then the Heart chamber vibrates like a crystal bowl triggered by trans-sensual creative passion always seeking new growth its next soul song! Indeed, the master code vibrates, oscillates, and calibrates, and sings the Heart's new soul-signature!

Understanding the consciousness and energy of this process allows you more awareness and tools for release applications. And this is the self-realization that the spirit, now embodied with you to gain all its wisdom for **new applications in the light.** This also allows you to know that during one of these upgrades or transformations, self-loving nurturance, remains the remedy for any symptoms of light-vessel transformation. This also allows your IAM spirit DNA master code to: monitor its light surgeries, transform you, comfort you, not react to mind fears, detox your human cells, and rewire your brain; so, you will be in the Grace of a regenerative free energy light-network vessel. Since the human form was only designed to handle fear, density, time-space, duality separation, and survival; the spirit IAM is so thankful for all the human-soul taught it. Thus, your Spirit Presence takes over and absorbs the human-soul-spirit experiences back into the Essence-core to be transformed into new wisdom-light applications and tools And, **only the Essence Heart core-code consciousness can change matter and energy** as you have all now realized. Again, Energy Light Masters, the new heart consciousness in the Divine-human light vessel transcends all physics, science, and technology. Indeed, this is **a new Human form. This is because you now fully embody the privilege to exist in your own consciousness; to exist in your very own love, and that is real success, real joy, real fulfillment. And this is real life mastery.** Heart Energy-Essence Master DNA code says; 'Only Heart essence can change programming or create new life. Heart Does no harm to self, other, or any part of life. All life evolution follows its natural code which nurtures and cares for all life. Heart is the spirit which guides IAM into new life.' With the Grace of allowing creation to do what is natural Essence coding, you realize **that fear** and all its negative thoughts, feelings, attitudes, and beliefs, when embraced by self-love and compassion; no longer want to be held in limitation, illusions, or separation communication. **They are just energy, and are asking to be set free to dissolve back into essence. So, your now aware that fear and love are both just energy. To experience All that is love and All that is not love, codifies as embodied soul's free will choice for 'All Creation' in the new prototype as the new species cosmic race-DNA codes.** And indeed, the heart vibrates and calibrates access to all light networks in the new light matter biology, not the human brain. **The brain** simply processes human-atomic data. Awareness replaces old human mind-emotions triggering meta sense Essences,

A NEW COSMIC RACE 129

because the wonderful human emotions merged with the soul's angel senses, and the spirit Essence, to create new light **matter imprints**. How about the Essence matter of chocolate covered compassion, tasty velvet pink sparkled joy, or green-lavender magnetic bubble shoes? How about the artistic-sense of climbing into your own painted vision of what your life can be, and there you are, inside that reality. These new meta-essence qualities of quantum-density, allow the atom and the quark-particles to dance in new creative delights. Now you have the imprint of Divine-Human, who can enjoy all new worlds of light, with New Earth school as a light-transport base and your own bio-light vessel for transport.

So, **do talk** to any residual fears, limitations, or illusions; as they await your Sovereign dominion and command that they too can be released into free energy essence to become something more. Yes, **fear wants to go home** to be embraced by love. Fear and love have taught and served each other, and you well, to experience all that is and is not love so you understand free will choice. You also understand full conscious responsibility for your own energy, and that only your essence heart can release Fear into free energy. And, you are reminded that you are first a Divine Being and always will be. Then, resistance is no longer a karmic pattern of having to fight self or other, or go against who you really are. Any fight or resistance was just the separation in SELF that said; "Someone will steal my love, power, or light if I **open to Self-Love**." Living in the moment of embodied Presence provides **just enough gravity** change to manifest heart's expressions and creations. It is **no longer resistance,** but the freedom to never have to go against SELF! This is because the light vessel heart acts as a magnetic-gravity anchor for light travel and living in moment-to-moment imprinted manifestations. It acts as a vacuum **graviton-warp drive** for your light vessel. Then you are **no longer**

**dependent on Gaia's light body** but have your own Light bio-vessel with your own regenerative elements.

And how have you **angels changed Creation**? This awareness, as human-angel-creator, gives profound experience to all 'All Creation' beyond just compassion; into new meta-sense joy, beauty, and giggles never known before. And yes, Divine Humans, have learned to grow the soul without harm to self, other, or any form of life. They have learned that **there is another way** through embodied experience without suffering, death, disease, or wound to embody an Essence Soul. The **embodied master-life angels** have told the cosmos they have experienced embodied love. Consciousness has grown soul's unique love coming to know Itself in infinite experiences through Its Creations. This love has grown human, angelic, and essence senses such that **the human experience of touch, a soft kiss, or the texture of rain becomes a new experience within light vessels 'quantum density experiences in new moment to moment awareness**. Earth is a school where one can learn soul love. And now Love embraces the human imperfection of its Spirit's perfection. And now you know the rest of the story. It's now all your own energy, so what do you wish to create with it? Your heart awareness will whisper that a decision made out of fear is not a choice, and that a meta-sense awareness is a free energy choice. It will also tell you that trust is just the Divine Heart transporting you back into your new consciousness. Sense the quantum density, blend/merge of all physical and non-physical meta-senses, to serve your instant creations now. Remember, **when you left creation to incept this universe**, you had the gene of compassion, but you did not yet realize this gene code would go on the long journey of growing its spirit essence in to angelic senses and then into human mind-emotions. This deep experience of allowing one's own soul, apart from the spiritual or soul group, to **be in love with all life experiences** of the: art, music, elemental matter, and human matters of life, was never experienced by the angelic worlds; before you grew this Divine-human DNA-species prototype. Hence, Heart consciousness always transcends all laws of physics, science, and technology. And now the cosmos will know **love beyond**: death, disease, suffering, loss. spiritual amnesia, and the fear of living. Conscious Love has witnessed and experienced the 'All That IS and The All That Isn't.' **Love gives, living-breathing-embodied authenticity, to the passion of Creation and Essence existence of Life Itself**. And even now, it is the very sacred chalice of the soul's newly born Heart and all its delicious potentials, that walks inside Love on ALL your worlds. Remember, these imprint potentials may first appear in the conscious awareness as symbols, images, metaphors, or codes before they manifest into reality. Therefore, the heart economy awareness is and equals, direct manifestation of experiences of joy, creativity, expression, or materiality, until they are simultaneous.

# Experiencing New Potentials-
## Awareness=Manifestation 8-2021

Embodied Life Masters, experiencing new Heart potentials is the fulfillment of creational evolution. Let's review the **energy dynamics of arising potentials and how new ones come to be**. Energy Potential Life Masters, your heart economy awareness, is and equals, direct manifestation of experiences of joy, creativity, expression, or materiality. However, these imprints may first appear in your awareness as symbols, codes, images, or metaphors until they are simultaneous. Here, awareness=instant manifestation. Your Divine-Human light vessel, avails your heart's biosphere open access of streaming consciousness, throughout the cosmos. As, **Energy Communication** Life Masters, you now have the awareness that the Soul of Creation exists in one moment of awareness inside your own experiential trans-sense potentials, inside your own heart chamber's creation. This is because you fully embodied, the gift and natural right to exist in your own consciousness and your soul's unique love. That is real success, real joy, real fulfillment, and real organic bio-life.

The soul thrives on new experiences. And, it comes from new integrated awareness. Again, these new DNA imprint potentials may first appear in the conscious awareness as symbols, images, metaphors, or codes before they manifest into matter reality. This preserves their **quantum density** such that the new human form in its **light vessel can recreate the sense of the touch of skin or a kiss, while also feeling its pure light-sense energy-essence at the same moment of direct experience**. This allows human emotion, angelic senses, and spirit essence to merge into new meta sense qualities, tones, and hues in the light vessel. The soul can use its upgraded light vessel as its instrument of experience instead of just the limited memory of human brain processor data experience. Indeed, you have all grown direct life experiences into super-senses, meta-awareness, and meta-potentials, in the unique Heart of the All That Is. And its Grace and Elegance continues to adapt and grow. Unique Heart's ultimate capacity using an evolving DNA-Essence master code; is to naturally allow the heart consciousness energy to go beyond the laws of physics, science, and technology. This gives you mastery over quanta bio-matter inside your own consciousness. Life Masters, you have self-realized that you are continually adapting and upgrading your light body to continue on in the New Earth realms and throughout the light meta universes.

Unique Hearts' Self-sovereign energy vessel of fluid potentials: lives, experiences, and light travels; where hearts seed particle or essence substance can change its experience of matter. This requires the flow of unconditional, unrestricted, energy flow throughout the Heart of Cosmos; and the evolving Essence experiences of all its soul-spirit creations. Creation is the constant movement of the energy of ITS consciousness. When energy is allowed to flow freely, there is no resistance, except by choice or perception. There is only: unconditional creation, unconditional love and unconditional freedom, no matter what, throughout all evolution cycles. There is no harm to self or any part of life when its Essence DNA life code is unrestricted. Your heart can then channel unlimited source energy flow through you, in direct moment to moment potentials, offering constant new experiences without recycled patterned living. The Heart of Creation inside your soul heart loves new experiences even if they seem like risky business! We reiterate, that what makes experience new is its awareness. **Its language,** in new DNA imprint potentials; may first appear in the conscious awareness as symbols, images, metaphors, or codes before they manifest into matter reality. Here awareness is/equals. instant manifestation.

As Divine energy Masters; your new consciousness has grown so deeply into the root of love's light essence core: that soon your food will remember to grow itself. Soon the salt in your oceans will purify your water like a giant crystal. Even now, your DNA remembers to transcribe you into bodies of light. Your sky will respond to your love and remember its essence light as quantum rainbow rays and particles, such that electronics will be outdated, and your trees can be your lampposts. **Light vessel has opened inter-stellar and cosmic multi-quantum senses, multi-time, and multi-potentials** in renewed perfection such that each of you can share in the 'All' of other's potentials in your upgraded life coding. This way you all ascend inside one beautiful New Earth Heart. This allows a Master of the Heart to say, "I Am in you and you are in me and we are each other's potentials, even in our soul's uniqueness."

Those of you **still integrating parallel selves** are now embodying the awareness that you are in constant multi-sense and multi-time space realizations of past, present, and future experiential potentials. Those that have already evolved into a one potential moment, one-life vessel, and one unique Presence are Master-Potential Creators ready for Soul's new awareness experiences. Soul Heart loves change and new experiences in the new Divine spark for the Aquarian Age Super-universes of light; and its transhuman superconscious natural abilities to play with life. Yes, this heart revolution offers the return of trans-human or super conscious senses and abilities. Now that you are ready to hold your own true consciousness value as trans-human, the freedom of your love's unrestricted energy can flow core essence light as the natural norm again. When you hold or image everything and everyone in their Source's natural essence, then humanity and every species on your planet will also remember Source love's growth code as creation. In that perfected re-connection, everyone and everything can be changed, just by allowing Source consciousness to make it so, as a **natural miracle of life. Your DNA remembers that severed limbs can regenerate, food can grow itself, and sound can levitate substances and synchronize your light vessel.** It also validates that it is impossible for Creation to neglect, abuse, control, or destroy its own creations despite the frightful or distorted conspiratorial creation stories your souls chose to live through. It also sends a strong message to Old Earth dinosaur systems that creating from light's love can be quite disruptive to dense matter realities that refuse change or are locked in separation of Self-patterns.

The perfection of energy consciousness itself, allows full expression, unencumbered by any judgment on its experience. The lover of life's energy is always renewed by its own love because it is committed to its own heart as its true beloved, only through direct unique soul experience of growth. So, what new-raw experience does your consciousness want, or do you consider it risky business? Again, what makes experience new, is its awareness. Here, there is freedom of souls' cosmic meta-sense expression; and an awareness that everything is their own energy, happening inside their own consciousness. Anything outside that is a participation in the illusion or theatre of another's soul's reality.

So, if you could remember that you could create out of the bio-matter substance of your own love as your own creator, would you need doctors, governments, technology, and space craft, as your reality? If your cells are made of the stars and universes, why do you need astrophysics? If your heart's biosphere is its own organism, why would you need science or physics? Is everything you need inside your own biosphere of light already there **waiting for you to aware yourself of it**? Can each room in your house walk you into anew universe or reality? Can you create your own inner-virtual world conscious realities? Can each expression be a new lifetime which has a different outcome, awareness's, and self-realizations? Your Divine spirit indulges passion in enjoying being human while simultaneously birthing many new inner worlds of momentary experiences with all their various new qualities, growth stages and light particles. *And, when that moment/creation is experienced, it dissolves back into free energy! Indeed, a moment is as a world born, experienced, enjoyed, transmitted throughout the cosmos, and released back into essence.* What does quantum kindness sound like? What does the sound of a light particle taste like? These Trans-senses offer new passion in continuous new consciousness streaming in a blend of multi-stellar simultaneous time, new angelic/divine senses, and moment to moment embodied potentials. These beautiful meta-sense heart experiences, with yourselves and the soul of Gaia, are pouring across your internet and communities across your world and can't be detoured. You are all going home to love as one Cosmic race of light in your new fancy standard of consciousness. Your arrival depends on each soul's awareness of their soul light's arrival into Life Mastery.

You're finally aware that your world's technocratic expressions have failed in their attempts to mimic or map this elegant Essence consciousness you have prototyped for the new human species with: replicated, implanted, augmented, virtual, hybrid robotic, and quantum chip realities. These energies are useful for developing technologies to heal all Old Earth Universal DNA races and evolve humanity; but do not have the consciousness to replicate or subvert your essence in-souled light creations. In fact, were they all not created from your own experiences of consciousness to understand your own energy and creations? However, in-souled essence bio-light vessel denies **any alien and unnatural bio-signal, in its biosphere.** Light body's prime progenitor angelic sense from essence core light-life code, encoded joy of creation in all its experiences; despite the acceptance of the human-soul experience of abuse programming inside separation. Indestructible core essence is eternally perfected in experiential love, casting out all source code information veiled and taught by fear. Such is the natural code for accepting life, and is the true nature of love's potential and activity in creation's existence. Hence, your soul's Eternal now allows all possible experience available to create and self-realize such that **you miss nothing** and have the expertise of all possible potentials in a physical or non-physical creational existence. However, the sovereign soul must aware that IT is ITS own energy and consciousness and that nothing outside that is real to its own creation; because the inner and outer core have merged and re-essence/d light.

This is true because light body returns all trapped energy into free energy, free will, and is a genetically sovereign vessel. Its consciousness allows all time as eternal now. Your past, present, and futures are all going on now; **thereby communicating/ merging/streaming into One Life moments of particle matter interactive potentials**. You're no longer living simultaneously in all these separated or parallel stories. In now potential, all those trillions of potentials were lived out through your atomic or matrix holographic universe; and their contiguous realizations are integrated via potential mastery of the meta/quantum universes. This is possible because you have grown a new vessel that is a Divine-Human with a re-essence/d soul merged into an ascended heart essence core light field. Your soul crystal and rainbow diamond heart have melted into liquid light. The bios-field lives its soul's life mastery: beyond the mind, beyond time-space, beyond physics, science, and technology; because its hearts vacuum chamber is now its own center of its gravity. This way you get to live every potential that has ever existed in the past, present, future; or that will exist, by allowing all your stories to magnetize back to the All of YOU that created them. We repeat, consciousness sets each experience free once its potential is fulfilled; so, you are never trapped or separated from your direct soul experience in any creation. However, Heart fulfillment will always follow new experiences and awareness inside your consciousness. It is the nature of the soul to grow through such rich, deep, bio-physical and quanta-particle density experience. Again, what makes experience new is its awareness.

In the **Old Earth holographic experience** of separation potentials were limited. For example, you can imagine what it would be like if you married another person instead of your partner and experienced all the movie versions of different outcomes that were possible. Now that one choice also changes all the permutations and combinations of all the potentials of both you and your partner and all the versions of this life and every other existence you have had in the cosmos with them or those his soul-spirit energetically interacted with! That also changes all the past, present, and future lifetimes you would live out and anyone in any future, past, or present lifetime, that existed in that lifetime with you. And what about all the alternate versions of your New Earths universes that might offer experience until your consciousness completes all possible expressions. Then you apply infinite outcomes to all the people on this planet and those you knew throughout the universe and in universes before and after Earth, and in the entire Cosmos. Now you understand that, your quantum-cosmic energy meta-senses, are unlimited in expression. Therefore, in the light vessel as a sovereign soul, **all those potentials are integrated into ONE new moment i**nside one potential with a new unlimited script. All that was in the future or past remains as Love's wisdom. Even the story, data, memory, dissolves as the meta sense self-realizes it was just for the experience to **have Creation live through You**? No mathematics, log rhythms, technology, science, or physics can compute an everchanging Eternal Presence of the All That is. Consciousness is always in the Love and delight of its creations via their direct experience of ITS everchanging energy awareness of cosmic potentials at all levels of existence using free energy to communicate with all life. Even those still in the awakening light integration of self-awareness, can enjoy seeding new multi-stellar or cosmic realities, once mastering linear time in this matrix universe; where they lived one story and all karmic/limited patterned potentials in sequential order. Then their heart's new light field can enjoy the new multi-quantum density senses, while stretching time-space gravity matter to enhance creating in new potential ways and live in the grace of life.

However, in this new enlightened consciousness, the heart's light vessel no longer creates from polarity which deals with force, control, power, and opposites in conflict always seeking resolution. Your Meta light Quantum fields go beyond time and use stellar/cosmic light and sonar/soma senses and trillions of constantly self-realizing potentials in bending and looping space and time to create new realities. You experience these in: light travels, tele-communications, new light-matter inventions, projecting your consciousness streaming into your technology, physics and science upgrades, meditative imagery, symbols, codes, and sensory guidance throughout your day, as multi-sense-time shifting reality potentials. These are inclusive of every kind of torsion wave conjugation or particle interaction. *Since the light spectrums of light, color, and sound are information life code carriers; you can now understand how your vessel alters by: looping, bending, stemming, particle-izing, imprinting, re-essence/ing, and fusing, plasma light potentiates for new matter biologics*. Your aware that the base line sense codes originated from a love joy, imagination, freedom, and beauty, to name a few, and have grown into star tingles or quantum-density sensations. What other senses of passion might you find in the trillions of new helix code patterns the cosmos wants to experience through you as your own cosmic creator? And remember, you can still visit the theaters of past-future lifetimes experiencing their potential outcomes or bring back all the information or emotional growth you learned in those lifetimes and use it in blending new multi-quantum senses. However, **you only visit** the theaters of the past/future. You, **no longer go back or stay**.

Now, imagine all the versions of reality being played out on your New Earth realms and super universe of light from all the representatives from all over the cosmos. And as Creators of the new Quantum Cosmic races, you remember and continue to transmit to those awakening; that Existence is Eternal, cannot be controlled, and its core essence cannot be extinguished. Your free energy heart knows that 'Its' own bio-sphere can dissolve into free energy or pure essence in any moment; such that an experience need never be repeated, stored, or memorized, or re-incarnated. And remember that, through the lens of the Heart of Creation, the only thing that Creation really cares about is experiencing your Essence Heart **with and through you.** And did you allow all the love you could hold and experience every possible potential your heart could create? Were you the tender care taker of your love, joy, and passion or did you expend all your energy on judging yourself and the cosmos for the experiences you chose and the very essence of You? Yes, only Essence matters and all matter is of Essence.

When humanity accepts and **allows themselves to love again**, their own natural unique Divine-human worth/value so deeply, even unto their core then: their trees will be as solar panels, their air will feed them light as food, they will naturally communicate with their cosmic neighbors; and create matter with their own consciousness without needing a, (computer/technology/science/physics), or anything outside their own soul consciousness. You simply imagine the perfect home and you're living in it! **And all technology will reset itself** to better serve your consciousness, or dissolve back into free energy. And on the New Earths food can grow itself, water can cleanse itself, and light bodies can light travel and regenerate their own DNA helixes to create without having any experience ever the same. How free and liquid is your new love masters, that all of the world can sense, be changed, and choose the social capital of joy's sharing in equality rather than the value of enslaved consumption. And that is why you have returned your perfection to this world and your perfection has returned to renew this world! There is no artistic matter in this world greater than such a radiant heart whose love can renew and re-perfect new life's dancing quantum symmetry. Love's Source Presence is growing ever-more trans-positional, abundant, healthy, and joyous in heart as your worlds return to the light realms and beyond. From flatline to flashpoint with one drop of essence you go! Do liberate your love deeper and deeper until all your other new worlds appear and you disappear into the light, never seen by the human's naked eye or even the eye of even artificial intelligence. Love yourselves, your core light, your hearts, and your consciousness so deeply that you are the miracle where consciousness lives in the dynamic of life's experiential potentials. Herein, there is only: unconditional creation, unconditional love and unconditional freedom, no matter what, throughout all evolution cycles. There is no harm to self or any part of life when its Essence DNA life code is unrestricted. What new unrestricted Heart Sense experiences are you ready to **COMMUNICATE with potential moment to potential moment?**

# Ascension Scenario Potentials
## Update 9-2021

Light Masters, here is an update of ascension and migration choices for those, just initiating their light codes, still awakening, or in the embodied self-realization progression. The bio-ascension, which will progress beyond the laws of physics, technology, and science for your new species bio-cell DNA; has been fully initiated, transmitted, and illuminated by all Quantum soul-master communication energy consciousness. This free energy communication opened the New Earth star gates for your local universe's trans-migrations. Its natural code presents as an: Essence Trans-human, Divine-human, or embodied living as a Sovereign Master energy communicator or Creator. These living master energy communicators have shifted their light bodies into the new code imprint-DNA Essence; or the free energy gem vessel of quantum particle interaction. Their full conscious meta-heart embodiment transmits the realization; that organic DNA bio-cell consciousness can change anything. However, these unique soul imprints may first appear in your core essence awareness as: symbols, codes, images, imprints, or metaphors. That is, until they are simultaneous into your hands and use, in joyful/playful of: creativity, expressions, experiences, and matter. **Here awareness=instant manifestation.**

An embodied ascendant sovereign Creator, (or conscious life master), can aware creations into instant manifestation, because the embodied soul has anchored itself in every part of matter that animates organic-essence existence/life. There is no judgment on self or any part of life, allowing raw free energy experience via awareness of total self-love/self-acceptance, through the DNA Heart code communication and guidance system for life's existence. Its currency is consciousness and its master communication energy in natural/organic trans-human genetics; with a base of 8 new rainbow-torsion fields for creating new quantum DNA helixes for quantum life. This proofing of consciousness illuminates an infinite unknown of, moment to moment potential choices and outcomes, for mass realities and all creations. This is the living master's gift of consciousness to the cosmos in accepting Life's Love as existence. Free energy cosmic communication masters, as Sovereign Beings, use their Heart's creation chamber as a unique mobile star gate, through which, all realms of creation inside their consciousness can be accessed. Quantum particles, excited by passion in the heart chamber, create stable quantum particle interactions, which create their reality moment to moment. Quantum interactions are always making love and creating from infinite unknowns, in an Eternal Self that breathes life into new expressive experiences. This natural creation heart pulse constantly moves through creational torus waves to support the changing cell consciousness of every life form in your world; while harvesting new existences, as Old Earth transforms into **multiple realms, realities, and new organic bio-spheres.**

This **embodied cellular Love** triggers/transfigures the new cosmic DNA quantum codes, along with Gaia, for humanity and all her evolutionary life forms; for unique soul choices for universal bio-enlightenment potentials around 2035 through 2077, depending on total light quotient consciousness. Bio-cell ascension is authenticated by the direct experience of the Divine-Self as the only Creator of its own soul's embodied reality. **Bio Ascendant upgrades, awareness, and self-realization** will continue to replace mind, emotions, gravity, time, polarity, power, energy and density. Embodied Self-realization will continue to replace thoughts, feelings, attitudes, and beliefs. Each unique Hearts' bios-field/sphere lives its soul's life mastery: beyond the mind, beyond time-space, beyond physics, and science, technology; because its hearts magnetic-vacuum chamber, is now its own center of its gravity. This is so, because your soul is already everything that has ever existed, even though you have allowed limited experiences of your own attributes and essence senses in your local universe to master space, time, density, duality, and an atomic body. However, they served your Essence to grow meta-sense core essence light attributes of quantum density; such a all life can communicate in higher frequency joy. Doesn't the tree light up with pink peach joy when you touch it or the book you wrote come alive like a movie for the reader.

All **essence souled groups,** just to example a few, will be guided or triggered via consciousness or light transmission and illumination. They include: embodied Multi-Rainbow/Communication Energy Masters, the Atlantis Crystal-Cyber energy Masters, the Millennial Masters, the Diamond-Rainbow Children, the Ascended Master Children, New World/Interstellar Multi-species Councils, or Special forces energy masters; as they manifest themselves into the New Earths and super-universe/s of light-love. They carry the multi-genetic diversity of the entire cosmos.

**All Old Earth Soul groups** will use any residual Old Earth Universal time-lines of distorted fractured past/future DNA codes to re-imprint to complete their soul evolution experiences of: polarity, time, density, gravity, mind, form, and cellular love. This allows their re-awakened hearts' core creation to fully open inside their vessels of light. As they awaken their dormant codes and protocols for the New Earths and light universes, their soul becomes aware of its choices to move to their next existence. This means everything held in a creation story of: judgment, doubt, blame, shame, death, disease, suffering, betrayal, abandonment, abuse, control, enslavement allows heart essence to: change, re-imprint, re-code, re-image the wound back into free energy, as rapidly as Humanity's sum consciousness will allow. When that moment is experienced, it dissolves back into free energy as an experience rather than trapped energy. **Then, a moment is as a world born,** experienced, enjoyed, transmitted throughout the cosmos, and released back into essence. And, **what makes experience new, is its awareness.** And all light beings have **grown trillions of essence-senses** since they incepted inside creation. In free energy vessel, there is freedom of soul's meta-sense expression; and an awareness that everything is the soul's own energy experiencing growth, happening inside its own consciousness. Anything outside that is a participation in the illusion or theatre of another's soul's reality. Souls have learned well they are responsible for their own energy and creations. Creation does not harm its creations or any part of life!

The **New Earth Star supernova shift** seals all past and future Universal time lines. Hence, the Eternal now cloaks and protects access for the embodied Master Heart gates that transmit the new DNA embodied-love codes from your genetic universe, for all the New Earths/light universes. It also marks planetary recognition of the birth of a new essence. It is an organic star-sun DNA or new Divine Human Species Being imprint; that each multi-universal over soul group came to master, through the base-12 DNA race-species constellations in your local universe. It is capable of **growing new helixes** for the next super-conscious generations of cosmic embodied Essence Human. This **Cosmic shift, initiated by the 2020 C-virus or DNA upgrade in consciousness,** happens individually and collectively, **until a coherent realization** rings a New Heart tone in the DNA harp throughout the All. This is a biological realization at all levels of existence that love and life have birthed a new cosmic day. The evolution of the atom and its mastery over matter transmits into the cosmic realms of the quantum-quark applications for new Creations. There is no Hierarchy here of creation, only differing streams of self-realized new particle-plasma consciousness expressing fulfillment of a journey inside evolution. As humanity remembers and takes responsibility for its role in creation, it translates itself back into natural creation consciousness. The wisdom of the Matrix Universe now lives inside each template for ascension, where each unique Essence Soul evolves their 'Own Universe', as their own creator. This ends the Matrix programs of control/power including fears of: the self, life/nothingness, human body, harm, abuse, or fear of one's own creations. Your Heart's Presence, in its natural state, is not afraid of itself, its creations, being hurt, or the bodies it creates. The Presence of life is not a slave of creation, since creation cannot be controlled. Energy is communication with all life. Spirit essence cannot be harmed by its own energy. *The **embodied Master** is* not afraid of their own IAM Essence or the forms it expresses in. These creators know they were never coded to be afraid of their own creations, afraid of being hurt, afraid of their bodies, afraid of their human, or live in a master-slave relationship. They sit in the passions of their own Creation where there is an endless supply of free energy to serve up new joy expressions. They become Self-love in Self-awareness in free energy transmitting and illuminating the gift of new life to all creation. Their bio-ascendant consciousness has transported back and forth using the new DNA bio-heart gates with the Old Earth ascended masters, since 2007, to open these new world energy corridors of consciousness. They transmit and become enlightened creation potentials using instant manifestation; via the quantum matter imprints of the New Earth Heart codes. This has created enhanced visionary potentials for embodied light living and triggers the use of the quantum creations as a grand tool for consciousness to accelerate humanity's self-awareness. This eases the transitions from the Old Earth Universe and authenticates living love. These Masters serve as living examples of a Creational Heart template. Increasing numbers of enlightened masters are now in self-realization inside their light-body network schools or group gatherings of consciousness, for your light's love revolution. These conscious light-networks are in tandem parallel with the Gaia's Heart Core, which is coded for light body ascension. Many from all over the world will now come forth to tell their stories of where they came from in the cosmos and their self-realizations about their own enlightened missions here. This brings actionable change through disclosure, consensus choices, unified consciousness, and new streaming conscious potentials to create with or change outcomes. As your planet reveals its true soul, amazing truths of who your species is and the star families you all come from will come to light. Then will come, the self-realizations, that your **star families are you in multi-diverse disguises of yourselves as the new cosmic race.** And, you have all have been in contact with yourselves all along and are ready to greet the rest of the cosmos with your new light-vessel Multi-helix DNA prototypes.

The **Multi-Dimensional Master consciousness** monitors and oversees the continued transmission of bio-embodiment and the conversations/communications of the cosmos from Old Earth to the multi-sphere New Earths. This allows these Light Masters streaming consciousness to over-light or mentor the Divine Human Star Seed and transmit the new creational helixes they trans-essence for the New-Earth realms. This includes transmissions that assist in the constant DNA upgrades needed to keep pace with the varying vibrations for new soul awakenings bio-enlightenment and bio-ascension. Their enlightened consciousness will also experience any Essence expression as their own manifestations living inside their own unique creational freedom. Since, the **average age on your planet**

**is 25,** this will allow these multi-masters consciousness to monitor, mentor, and help stabilize the light bearers and children, until they are fully bio-conscious. They also may choose to help oversee the New Earth-Star family cosmic interstellar communications; since most have great experiential wisdom from light universes for the massive re-education programs that will be required. All bio-vessels across the planet are being rebooted daily to receive these contact transmissions and assistance: one on one, in small groups, and on mass according to chosen consensus realities.

The **Millennial masters**, many of who were embodied in Atlantis and remember the misuse of energy have returned. They are busy changing the consciousness for benevolent systems such as: eco-sphere environments, cyber magna-tronics, astro-sonics, citizen world councils, monetary and exchange equity, conscious health systems, living matter housing, and restoring women and children's Divine rights. Their **gaming platforms** are providing new organic cyber-light networks. Thousands of these light masters come together in virtual light networks in the new sport of gaming which is also bringing massive benevolent innovations for: the virtual worlds, science, technology, and physics and beyond. They trigger each other's code to activate their core light biology and heart channel communications. On mass, they release the new *cryptology* for benevolent artificial intelligence via the new source code creation helixes. Their soul's code imprints create solutions, problem solving, and re-education protocols outside the matrix. Each game platform can serve as a prototype for massive re-education on any subject matter including the bio-light vessel, and the new quantum consciousness applications for all aspects of trans-human living. This consciousness will dissolve or over-light the old dinosaur applications to your changing Universe. Their consciousness will also prevent social, cyber, or digital, dictatorships and AI (artificial intelligence) systems from *subverting* the ascension for GAIA and all her New Earth realms.

There will be inter-world or stellar **ambassadors** for those who wish to work with the Hybrid-ET programs, if this scenario is chosen by world consciousness. Both artificial intelligence-ET hybrids; and the hybrid-human DNA programs are viable global careers for creating new species life-styles and allow all past/future universal races to heal and understand soul essence heart and its original gene of compassion. The hybrid-ET programs can help re-colonize the Old Earths for healing and serve the Ultra-light Essence bio-ascension migration to the New Earth realms.

The **Inner Earth consciousness/councils** work with inter-species civilizations, as well as star family contact teams for the many soul species who are being transported to the New Earths. This includes the animal, crystal, plant, and elemental kingdoms. Indeed, **your oceans** have more new species to grow new life than all the stars in the cosmos! There is ongoing re-education and reset in every area of life, weather in the transition preparation on the Old Earth to reset for the New Earths galaxies and systems. You have all created **transition realities** such as: soul trauma centers, life transition teams, children's light centers, medical-DNA centers, and food domes. There will also be bio-sphere and bio-plasma ship communication teams to assist with bio-plasma dynamics, since the light vessel is a free energy vessel. There are also inter-species language translators for the star family introduction teams channeling new light communication or ending karmic life agreements. *All versions of awakened or unawaken soul reality will be healed or set free into higher consciousness. Take notice of how light fusion is fracturing space-time as all past-future time lines/existences heal and converge in new reality experiences offered by the light universes. Its language has already entered your new* *careers and lifestyles as in: designer imprint/replication technologies, plasma magnetics, Astro-sonics, interstellar commerce, intra-world or stellar inter-species sociocultural communication networks; cyber-sonic spaceport systems, or lifestyle mentoring and fulfillment centers. These are some of the careers the light children will pursue in their light vessel generations.*

The **Cyber-Technology master consciousness** is widespread and distributed world-wide while the old Earth Electric information networks and cyber networks short circuit while being transitioned into **Magnetic transponder technologies** with the suns, the moons, and geothermal energies. The new multi-Universal communication networks, made of organic bio-sensors, are **living conscious networks** naturally calibrated to the Divine Human heart gates. So, the light vessel **Essence DNA heart-gate cell** acts as your: transporter star gate, a magnetic imprinter, Source Code/r, centrifuge, quark stem cell particle and bio-ship for New Earth spirit matter, inside embodied love?

All Lighted Souls are already using multi-soma conscious awareness. In an instant they stream unity consciousness to: comprehend parallel realities or new realms, transmit solutions to end planetary abuses, connect to free energy systems, or exchange new types of valued equity. These Light Masters lives remind humanity how; imagination inside a self-loving and life-loving consciousness brings instant materialization and core soul-IAM Essence expression, in living love codes of the infinite unknowns. Hence, quantum doses of light will continue to amplify **humanities choices** to disconnect all electrical neurological and brain systems of Old Earth. This includes: old DNA ancestor memory, old bodies, pasts, futures, and massive outdated brain data network memories. It also includes old thoughts, beliefs, attitudes, stories, and perceptions from any time space density ever experienced in your local universe. We reassure that the New Spirit of Earth has already birthed New Multi Earth light plasma bio-Spheres within the Super Universe. Indeed, Gaia will experience the New Essences of all her life forms including her minerals, plants, and animal kingdoms. This also means Gaia's core light has already birthed and formatted anew bio-light imprint over her own Old Earth Body creating her new realms. This allows Gaia's old matrix hologram realities to release, so her bio-vessel creates its own re-genesis into new bio-spheres, allowing her to act as a space-port for light vessel transport.

As humanity releases and heals Old Earth linear time lines of human futures and pasts, transport through the heart gate to parallel New Earths, **changes the face of death**. Conscious Bio-soul embodiment will replace the death reincarnation cycle. Death and birth will once again be understood as a natural light transfer of a soul's changing evolution and form, rather than enslavement-punishment or torture for failure of unworthy living. **The light transport or migration of soul's essence can now appear in many forms** as each soul creates a heart chamber suitable to match their soul's composite evolution. **Some souls will essence that:** they have been lifted, beamed, disappeared into the light; or walked into the aurora borealis, entered heaven, or were transported by a starship or angels. Some may experience being slip-streamed thru a star gate or molted into changeable body forms. Others sense themselves as re-particle/d light. Some create being rescued and reunited with their star family, masters, angels, cosmic beings; or even having entered a new intra-world school. Others may transmigrate to higher New Earth realms so as to act as a guide to their Earth families. *A handful might ascend through their own star gate where they enter the Super universe. Inside each core soul light consciousness, lives experience in a new quantum frontier of expressive potentials with new cosmic creations* *or universes which the cosmos has never conjured before.* Yet, others might experience themselves as a part of A Federation of Worlds and engage New Earths in new contacts with all other star worlds and interplanetary systems. Endless soul choices abound as the light body will continue to offer new ways of coming and going from new Earth realms/realities not yet realized in your new vessels over the next Aquarian light cycle. And Humanity chose your C-virus of 2020 to activate every DNA on your world to initiate this new cosmic cycle. Humanity chose to change outcomes and beyond cataclysm, world war, or alien invasion to awaken new DNA-potentials for their cosmic cycle/age of light, as the quantum heart energies accelerate old karmic patterns locked in dense polarity human body mind. This has made room for the divine Self to open the dormant DNA codes.

We reassure, that all **versions of transitional New Earth access** are viable under the natural parameters of cosmic bio-quantum dynamics and each soul's consciousness evolution codes and templates. All this is ongoing as New Earth plays the theatrical role of the new school for soul-embodied quantum light or multi-verse light training. Cosmic citizenry, free will bio-embodiment, sovereign creator, solar language coding, and any course experience of soul evolution you can imagine are offered. All of these scenarios will be played out in each soul's unique consciousness until each soul migrates out of this universe to its next existence. **Soul's light** will realize that what is in one's consciousness is their only reality, even within all the New Earth realms or biospheres. New plasma matter trans-physics will enhance the understanding of these realities. You will continue to stream forth love's potentials networking through your global quanta matter experience and streams of consciousness. This allows for the realization that there are New Earth biospheres within biospheres, realms within universes that you created before you came to Earth; and new ones awaiting you in the super universe of light. And now you're traveling with Earth-Gaia into her new homes in the **cosmos to self-realize them.** This has allowed you to become the new Source Code in Cosmic Heart's Petri dish for Spirit's new Bio-Quanta creational worlds. And yes, you have mastered spirit in space-time atomic thought matter. This is valuable knowledge to the rest of the cosmos that knows little of physical matter embodiment of the trans-sensual organic Divine Human. This will provide the consciousness for humanity, when each soul is ready, to remember how to master quantum Essence matter in order to join the rest of your cosmic neighborhood.

So, imagine translocation in your **light body/vessel** to: a new realm, or an avatar **version of a New Earth biosphere;** or even exchanging essence with other essence light or cosmic beings. Perhaps, you simply illuminate through each other's essence to communicate. Your biosphere might sense as a paradise of living Essence in rainbows of light. *Here, dreams are replaced by imagination. Imagination is real. Essence is real. The Essence is the vessel. Awareness is manifestation. The Essence is its own inner technology. The plants, animals and crystals commune with all life.* Most light body creations appear as more evolved versions of one's own soul essence in a trans-sense blend of new/old earth wisdom experience. However, some are indigenous to the new creation; just as the humans on Earth were once indigenous before their natural evolutionary cycle went into doubt of judgment separation millennia ago. All life is free to interact in loving harmony. Even the elemental Devas and the faerie of this world who attend to their creations are free to play. The water might taste like sparkling colors of liquefied light minerals. Charged water on the Old Earth had an electrical or vacuum spin. Light body essence cells can manufacture or drink magnetized water by heart's mineral light. You can eat droplets of essence or absorb an energy imprint of the taste of food without needing to consume its core essence. Essence/ing has to do with sharing a deep cell connection/communication energy with realms of nature that have also evolved through their own creation code symmetry. You can create the touch of human skin or a kiss, while also sensing its essence energy. The skin shade changes slightly to suit the quantum light spectrums and the moment's expression or experience desired. Or, the entire Essence can morph or re-imprint as any life form. Also, the body light feels so transparent, so effortless, and so attuned to receiving input as essence qualities; that no experience needs memory or learning. These transparent senses replace outdated input systems such as dense touch or mind-words. Energy Communication comes through attributes or streaming awareness of multi-layered or meta-sense color, light, and sounds. Nothing is absolutely solid and changes as the vessel uses its core light's excitement. Again, imagination here is real and replaces the need to dream because the Heart passion is its own manifestation of the essence you want to be.

The **essence vessel** sits Present to all life and allowing any heart passion, moment to moment potential to potential that arises to re-imprint or create life. Hence, you can create any version of New Earth existence you wish to experience as a creational embodied life master and enjoy any of the trillions of scenarios and choices soul-light living now avails! Consciousness is always in the Love and delight of its creations via direct experience of ITS everchanging energy communication and awareness of cosmic potentials at all levels of existence. Remember **to call in the awareness,** that you All Are/I AM, the transfiguration Embodied Love, manifesting into the Hearts and Minds of all humanity now. *Enjoy all the new experiential potentials in your next cycle of cosmic evolution as you explore how your new cosmic race and all your: crystal, plant, animal, and hybrid species live in the light. What clothes, homes, lifestyles, foods, light vessels, technologies, relationships, families, and new interstellar life systems you will design?*

# What did Cosmic Heart Essence Change? 10/2021

Energy Communication Masters, what has your Cosmic Heart Essence done? Have humanity's tears opened the global hearts enough to change awareness; such that your transmission of meta-senses in the new human light form can allow the entire cosmic reset another quantum leap in 2022 and even surprise love itself?

Enter Aquarian Age. How much did Humanity change when their new viral consciousness activated their Core crystal/diamond cell DNA? How much changed as their Essence Hearts grieved all their mis-qualified energies in being seduced by its fear, control, and mental programming; as if creation and energy itself could be forced or hijacked. Indeed, has humanity stopped abusing and suffering in love along with the worlds within them? Did they remember that they were first Plasma-Sun Divine Beings, and it was a journey into existence to master time-space density wearing an angelic human soul? Did they release distorted energies, false dreams, and addictions to suffering, death, and disease; or are they still under its hypnotic illusion? Did they remember that they are first and ever Divine beings and can't lose the innate organic essence bonding code of their DNA stems' bio-love? Did they replace service and suffering with all that life can offer from an open heart? Did they give up hope of the promise they have yet to live as they migrated to their next level of vibration or existence? Are their minds still torturing them or did they finally listen inside their hearts to their core Essence Presence whispering in their ear; that it will caretake, nurture, and love them if they embody life's existence, with no need to validate or justify, except to experience it all. Are they in a light revolution of change; where they will no longer give their love and worth away to the viral energies of power agendas and controllers, who would barter their right to exist by selling them fear? *Has it has come as the need to validate their equal right to the unlimited resources of creation offered by the freedom of the awareness of their own unique soul creations?* Will they channel all this into creative solutions or still riot at what was withheld; and being deceived by the gaming systems they blame for falling into illusion? Who broke their essence bond to creation?

More than any time in the history of your local universe humanity continues sensing the: symbols, images, codes, videos, stories, or internal aspects of the multi-holographic layers of everything that it created. And now, will unconscious grief and anguish remain the underbelly of its violence, anger, blame, shame, hurt, abandonment, doubt, judgment, regret that were never resolved in fearing each soul's own energies and experiences? These were the unresolved trapped energies in your universe of spirit essence and angelic senses that later densified into negative human emotions. Did humanity's love and light expose them? The opportunity to go inside, be still, connect with loved ones and be responsible for one's creations and energy has offered incredible opportunity to realign Soul purpose across your Universe. The limited human mind still clamors for the old safe predictable life, the old lifetime identities, past/future life times, wounded patterns, old spirit families, old limitations. The mind rages fear but projects a false safety to keep control of the human; for that is all it knows! **Mind will:** throw temper tantrums, anxiety attacks, fight with its own selves, act out addictions, or go insane; thinking it's keeping the brain and body safe from death. **The meta-sense heart of the light body knows** what it needs to self-sustain its own existence, as it is natural in the essence DNA code to follow this. It was the past suffering of a distorted creational story where competition and separation became a suffering mind that doubted the heart of Creation's bond to Essence;

rather than the gifts of receiving all that creation has to offer. Humanity had come to accept limitation of every form of fear as a prison of trapped distorted energies they would never escape and trapped in a body that lost its soul's self-healing biological organism. **How many hearts** opened and allowed the DNA bio-cell transformation of soul's core pattern to dissolve and re-imprint heart's new life codes? How many tears cleansed the wounded core pattern of: the right to exist without having to validate yourself, disappear, hide your light, prove worth-value, pretend; or empath for other's suffering in the constant dying of the old body, wounded lifetimes, fragmented DNA, unable to trust a heart that seemed always searching for fulfillment that was unattainable. Did the new **heart take off its mask** and awaken and say no more to anything outside its nature? Or, are the many still crying out to the heart of their spirit; "Please don't let this mind game and its nasty limitations blow me up, make me insane, or keep the polarity arguments torturing me like a robot, before I manifest/self-realize my most fulfilling potentials! And maybe I might even meet my true self in a burst of light?" Did humanity bounce off enough polarities to finally collapse into their core light? Did your world in all that time alone with family, with Self, without work, without distractions; face self without a mask? Did they reset their priorities*? Did they realize that real abundance, real manifestation is self-value, self-acceptance, self-awareness, and self-trust of their spirit? Does each soul steward their own love?* Did humanity's soul indwell in a vessel of light they created for it by mastering all human space-time experience in dark/light nights of their soul, which was recording every experience along the journey? And was there a hush all over the world where, humanity's quite spirits seemed so happy in the Presence of stillness; to reset their hearts journey and review all the growth, movies, collective chaos, and wisdom since it was separated from its soul and human aspects. And didn't Spirit even offer its Soul a new heart biology of light to house its full Presence of light's quantum spectrum. Humanity had to experience a **soul-merge** of one or more aspects of self, even though they weren't sure who was sharing the body with them other than the mind. And now to consider life from having a soul or loved ones who did? Now what was life all about and was there some master code that bonded one to creation in some mysterious formula for life? Did everyone have it or was it just for secret governments, aliens and gods, or AI special humans? Who had the power to the secret sauce of science, technology, or physics or even beyond all three; that would give you all the treasures only stored in heaven? What did humanity discover? Did their heart consciousness have the key or was that too simple? *Was awareness in the heart enough to change the laws of science, physics, and technology or were there going to be decades more of slogging out realities; until humanity realized that growing their unique souls in a cosmic petri-dish would transform creation. And did this involve an evolved human soul in a new Divine human form?*

And what did you **Light Masters transmit/communicate to humanity** through your streaming consciousness during the planetary reset? What did your hearts change? Even for many of you trusting your Spirit's Presence remained a stretch as you fluctuated your multi-variant frequencies inside the physical and non-physical light spectrums. And just when you felt you fully integrated your light body, another aspect in the higher Old Earth realms of light who had never experienced density, needed to share its movies to be bio-merged. So, juggling all these multiple-life aspects calling out for integration in your light body to dissolve polarities was consuming; and many of these aspects were never fully embodied before. And even now some master lights may still try to hang on to the old human form that the mind is terrified to allow to transform. Indeed, allow any distorted memory pattern in every cell as it cries and dissolves; so that in the new vessel, Spirit's love in the cell is sense-bonded in the DNA/RNA master code. *Aren't your teachers of experience, fear and love, just energy seeking freedom back to essence? Then, <u>Heart awareness returns</u>. Then, Heart Awareness frequency, is/and equals, instant manifestation of heart's joy, fulfillment potential and creativity as the natural bond to your own creation; which as a divine being you always had*. Bonded in a composite true stable light quotient or frequency of your own Soul and Spirit IAM, while embodied in a new vessel or quantum density, allows all potentials direct experience of their new essence flavors. Here, you know the result is that the heart's quantum essence-density can change anything is its own creation chamber, and enjoy blending attribute essences to create new homes, technology imprints, or organic matter. Now awareness is and = manifestation. *Whatever the heart is aware of in expression, joy of life, materiality, creativity, manifests directly out of Trust and communication of your own inner connection and the inner bond to your Heart's excitement or inner chamber stillness*. This allows your spirit to reveal each upgrade step by step as the old vessel gradually re-particles into the new model light body. In its embodying of **the bio-instrument** that you have prepared for Spirit by releasing the Old Earth Universe, there is no need for validation of your existence over and over in lost identity, purpose, or lost connection; for that would make you a victim of creation itself. That would just be another distortion. That is why you allowed yourself to experience life review after life review, as you went back and forth to your version of new Earth and reviewed the movies of your life on the old earth simultaneously. This **tele-transport in frequencies** opened your true composite essence vibration to remember your first light as natural essence. You remember that you don't need to control, power up, or manipulate anyone or anything to allow yourself to exist and receive new life. And that life is you because you have mastered all life. That's natural evolution as star seeds and midwives of worlds, seeking always to evolve creation codes to spawn and secure new life. Your energy is free and you're the everything of your own creation. In any master DNA life code, it is unnatural to steal or bargain your soul to access creational energy communication. It is also unnatural to carry the dense gravity of ancient ancestral DNA cellular grief from all prior generations.

**Gravity in your light** comes by allowing the Heart and all those feelings to arise in your awareness so they are forever: released, dissolved, re-molecule/d, re-essence/ed, re-imprinted, or changed by the natural heart code. This composite Heart with a bigger light capacity and frequency to create magnetizes awareness across your world so much that food can grow itself. The DNA can re-splice itself. The rain can purify any toxins, and your vessels simply snack on the energy of pink turquoise joy. It makes a new space to be filled by heart awareness gifts and the potential of fulfillment in any moment Heart's light matter imprints a quantum sense. How about tasting pink chocolate covered compassion or tasty velvet pink sparkled joy? Can you jump realities in green-lavender magnetic space boots; or imagine objects that float to you? How about the artistic-sense of climbing into your own painted vision of what your life can be, and there you are, inside that **reality in a virtual moment?** And indeed, it cancels all other realities. These new meta-essence qualities of quantum-density, allow the atom and the quark-particles to dance in new creative delights. Now you have the imprint of Divine-Human, who can enjoy all new worlds of light, with New Earth school as a light-transport base and your own bio–light vessel for transport.

*This Creation code ensures a divine human unique soul will fulfill all its potentials.* So did your heart lift the quarantine off your creations, to knowing, that all will be healed as the tears are wiped away. Did your salt-crystal tears tell that the heart of creation lives inside all souls and its heart is the only way in or out. It is the only freedom? Energy Communication Masters, in the New Earth Chronicles of the light body, you have finally arrived in your own true reality and in anew human form. This is measured by the alpha light-body constant or your unique Heart signature's creation code. This is your atom's soul mass to light ratio, measured parallel to your unique new spirit's light to mass quantum ratio. All you're Stories, all your conscious dreaming, or those you meet when you pretend to be asleep; or in stasis from this reality are in potential realizations of their experiences. They can all remain **versions of you** until you collect all your selves into one new Master Self. All your poems, your holography of self, your perceptions, and your awareness take on multi-self-realizations in the continuum. That remains until energy communication in your aware trans-sensual Divine-human knows, that nothing else matters, and that it's Spirit will never let it go without. **Then whatever is in the heart awareness, IS!** Energy Communication Masters, now you know <u>why</u> the new heart consciousness in the Divine-human light vessel can transcend all physics, science, and technology. *It is because it has evolved a new consciousness in a new Human form to embody soul to exist in its very own consciousness; to exist in its responsibility to its own energy and own love. And that, is real success, real joy, real fulfillment; and real-life mastery.* <u>The simple key</u> is a super-conscious or meta-sense Heart with elegant quantum senses, which can sense any energy and change anything in a moment of essence awareness! **Hearts very awareness is its own Source manifestation and** ready to take humanity beyond physics, science, and technology. So, is the heart of humanity ready to change their awareness, that openly; such that the heart is; their own greenhouse, doctor, loving partner, technology, or new galaxy? What did your heart change today? *What reality did it switch to Human, Divine, or walk in and out of both at the same moment...or go beyond?*

# Accepting Self and All Life in
# Age of Light 11/12-2021

Energy masters, In the eyes and the heart of Creation all experiences are equal. **Accepting Self and all life is how life is to be lived in the age of light?** Only a perception of judgement, that life is limited in any way, could cause an experience to be separate from self or any part of life. What is life with Judgment? With judgment, Heart essence communication would then have an experience that would limit or lock out natural manifestations. Judgment would inhibit the free flow of the energy of experiences of both positive and negative: thoughts, feelings, attitudes, and beliefs; to neutralize any harm to self or life. Judgment would then lock out new experiences of life's fulfillment potentials. Judgment further misreads that soul is not first a Divine being having human soul adventures. What if you had an experience with an awareness of judgment? Then its underbelly of addictive mind, which recycles through the old earth energy collective unconscious of; anger, blame, shame, doubt, regret, suffering, death, and disease would find its way back to its natural essence heart to be freed. Thereby, life's love could repurpose the soul moment to moment. Your own energy communication would release any stuck/trapped energy by embracing everyone and everything as equal in experience? Could you then enjoy each experience in fulfilment, growth, teaching, and even in all quantum sense densities at once? Are love and fear just energies of experience that are ready to go beyond themselves into new essence awareness, *which can sense any energy, and change anything in a moment's potential? Then whatever is in the heart awareness, IS!*

In accepting Self and All life in living potentials your meta-awareness moves polarity: fat/skinny, worthy/ unworthy, lack/ abundance, bad/good, light/dark moves into integrated blended senses and multi-diversity of experience, rather than recreating any wound or harm. Do you notice when you communicate in relationship with life with what you: eat, sleep, drink, take medicine, relate to people, or live without fulfillment or passion? Does your aware self still attempt to accept any experience where heart senses no potential of sensing new qualities of natural joy? Have you shifted to where awareness manifests as only experiences of full-on living? Is your creational heart embodied, in its new consciousness and in its new Human-soul light vessel, ready to exist in its very own expressions and evolving experiences of love? Would that authenticate natural abundance, real joy, real fulfillment; and real-life mastery? Do you pulse with the heart's magnetic pull in acceptance flow of new trans-sense experience knowing the essence heart can't be illusioned by natural energies outside its essence communication with all life? Indeed, organic essence life always follows its natural life-code and embraces soul's newly grown essence senses. as your own intimate relationship with life. Exploring heart's passion potentials, in your energy with self-loving acceptance communicates how to live in the next cycle of light? The embodied **ascendant sovereign Creator** *consciousness* can aware itself into manifestation, because the embodied soul has anchored itself in every part of matter that animates organic-essence. They know that direct experience via awareness or self-love realization through the DNA Heart code is the only true guidance system for life's existence.

# Quantum Attributes in Manifesting Awareness 1-2022

Energy Communication Masters, now you know why the new consciousness in your Divine-Human light vessel can transcend all physics, science, and technology. It is because it has evolved a new consciousness in a new Human form to embody soul and to exist in its very own expressions of love. And that, is real success, real joy, real fulfillment; and real-life mastery. The simple key is a super-conscious or meta-sense Heart with elegant quantum senses, **which can sense any energy and change anything in a moment of essence awareness!** <u>**Then whatever is in the heart awareness, IS!**</u>

**Hearts very awareness is its own Source manifestation and** ready to take humanity beyond physics, science, and technology. Your planet is operating **the light transport base for the new cosmic race** light vessel prototype. Such is free energy in full out living, as essence heart communication within soul's living mastery of creation. So, is the heart of humanity **ready to manifest their new quantum attributes into awareness**, so freely; such that the heart is their own greenhouse, doctor, loving partner, technology, or new galaxy? And how many are ready to **switch reality back and forth** between human and Divine, or walk in and out of both at the same moment... or go beyond?

Your planet operating as a kind of cosmic spaceport, is still undergoing its <u>re-education by the light</u> or re-genesis, through the **<u>repurpose of Essence</u>** in every area of life, including the ecology of relationships with: people, places, events, systems; as well as the animal, plant, crystal and all elements. This is the base for reformatting all your life systems in the light and allowing **your soul to be the living art forms they were meant to be**. In the Old Earth Universe extreme polarity reduced relationships to programmed experiences forcing and mimicking creation through false essence attributes of mind or technology, using illusions such as power, control, feeding addictive or tyrant energies. It offered the illusion that love comes through a hypnotic-addictive mind, instead of the heart's **essence attributes** of natural grace, joy, compassion, innocence, and an excitement for creating. These storied illusions have been a kind of spiritual propaganda of versions of truth, amnesia games of not being who you really are. Illusions offered' Self 'against its own heart accepting death, disease, or suffering to replace your natural essence sense attributes. Your creative arts have been your life lines to organic memory and expression of your love's aspects and their expressive attributes. But even these Experiences became dense reincarnated identities rather than allowing all the attributes to manifest themselves through your unique soul aspects **to inform your universe** of what its love was becoming.

However, you have realized that you allowed all illusions of experience to serve you as **a contrast, not a polarity,** of all the different essence layers of the light spectrum rays or energy bands that built your Creation. These were to express your **souls as living art forms**. Each ray or plasma band were really streaming consciousness and you called them the angelic realms. You were these realms and light ray bands as magnificent light languages. These were available potentials inside DNA essence attributes or codes of light, color, sound frequencies for each layer or dimension of soma-senses or streaming consciousness throughout the universe. Like **artists,** you created each layer in your universe in order **to** essence all aspect attributes in full ranges of expression in both atomic and quantum spheres/worlds. For example, the color blue has manifested itself and all its DNA ray attributes through you as its **life carrier** such as a: soul light, as sadness, softness, a blue spruce tree, cold fusion particle light, free will, magnetic lazar vector, heart emergence, or even a cobalt blue crayon, or a sky. And yes, **you are these attributes and these attributes are you**. So, you as the creator of these attributes become the creation and the created simultaneously. Hence, as a life wave carrier you're enhancing your own soul and spirit's creations through evolving your natural organic senses to become **their own manifestations**. This is how you inform the universe of what it is in its own becoming. Each attribute such as: grace, love, joy, innocence, elegance, freedom, transfiguration, or infinite light are in their becoming and their own realization to inform the cosmos in its continual becoming and realization of 'ITSELF.' This allows your Eternal Self to become its own manifestation and is the ultimate freedom and abundance. *Hence each attribute is like a new flower, a new organism, a new helix which informs the cosmos what is becoming through its own creations* and its growing essence attribute expressions. The creations of new Essence attributes are limitless. How many humans, soul, or spirit **lifetimes** have you created to enjoy yourselves in **different aspects of these attributes** to explore creation?

Each Energy Communication Master allow the creation heart to live its essences of: beauty, joy, innocence, grace, free will, creativity and endless potential attributes yet to be experienced, in their own becoming. The heart communicates these attributes moment to moment and potential to potential.

**There is no more waiting**! In the past many of you have been frustrated because when others wouldn't receive your love or manifestation of these magnificent attributes. You thought you couldn't give them to yourselves; holding back your own light's love. This made you question your purpose and existence because you felt you couldn't fulfill your natural organic joy. This old misperception set the pace for looking outside you for love instead of **purposing your unique essence**. Your consciousness created existing life for the purpose of growing creational potentials in a soul light vessel that could manifest Divine human meta-sense attributes, which was **your true purpose all along.** Your consciousness said to you that; "When you're Heart's Being-ness in its awareness manifests these new Quantum attributes; then they are their own unique encrypted soul code, as and through you, at every level of existence. This universe offered the All in All an opportunity to use your own soul's evolution to have joy self-realize what it is and how it manifests and how it transmits/informs continued arising potentials. This forms a beautiful new Mandela of what joy and all other attributes mean to each soul, so creation enjoys it awareness of what 'IT IS.' This allows each Source Creation to share essence and grow pools of greater diversity for all life.

A simple joy of touching a snowflake, emulates the journey of essence life descending from pure consciousness into an energy of existence. It's journey into plasma gases, radiation, light water, atmospheric condensation, water drops, is/was constantly changing itself to become the experience of its perfect code symmetry in creation. To melt upon the Earth, it had to become or interact with all the base fractal elements of creation. Everything comes and goes from consciousness. The attributes of the atomic elements have offered a base for the ability **to purpose heart's essence** and build your quantum instrument to enjoy new attributes **in** direct manifestation of experiences of joy, creativity, expression, or materiality. These attributes grew from becoming all aspects or archetypal experiences offered in your bio-physical universe. This includes all the aspects of lifetimes and their attributes. You've have been negative aspects such as: blame, shame, judgment, pain, as well as elegant joy and playful selves. You have been the mothers, sons, daughters, lovers, children of creation. You have been doctors, priests, warriors, space commanders to unique your soul in life. Each experience offered new aspects with a multiplicity of attributes to master polarized time-space inside free will. You even grew a human form with: mental-emotional-physical-spiritual animations to accommodate any separation, distortion, or density from your essence's consciousness. You have played all roles with your soul and over soul groups to master evolution of your soul's essence. And these new attributes of quantum density have arrived; such that whatever heart is aware of, IS!

These dormant **new attributes** were re-birthed and through your 12 orbs, which replace the old Earth 12 chakra system, completing the Old Earth matrix universe. Your new **13th orb, as Sovereign Creator, inside a conscious heart,** folds all the 12 orbs together into your **new ascendant heart's** (free energy plasma-particle light), or quantum-torus vessel. This is akin to mixing and matching the atomic rainbow light spectrum with the quantum spectrum rainbow to create new torsion rainbow spheres. Your unique new torus biosphere, can then form as many magnetic rainbow torus fields as you which, to explore all the new attributes and exotic particles for the New Earth universes and their new helixes.

**Your** heart functions as stargate, magnetic transponder, regeneration chamber, and graviton field with new essence attributes potentials its only purpose to fulfill. Each attribute informs the others and with any combinations of the attributes and their meta-sense qualities new life creates. Aspects and attributes of your atomic rainbow fields mixed with the quantum quark rainbow fields provide organic_combinations to produce new light life systems in beautiful **multi tones, hues, and meta-senses**. This could be called **interstellar matter biologics or bio-genesis**. For **example,** all

**shades** of yellow inside compassion can help build designer homes for those homeless from natural disasters. Shades of _**rose-pinks**_ are inside tenderness or playfulness to help children feel loved and supported in their soul's gifts and their Earth walk. Attributes of _**blue**_ inside the sense qualities of empowerment or emergence, can produce lazar blue inoculants and particle light inventions. All _**shades of white**_ create qualities of transparency and new essence consciousness to reform your world's governing by its citizens. Its clarity shade offers interstellar communication systems. _**Greens**_ are inside qualities of honesty and equal value to bring the justice of the soul for access to equitable self-sustaining life systems. It can also balance cyber-sonic frequencies. _**Lilac violets**_ have all the qualities of release and unconditional freedom, so all life and species are valued in equity. It also serves in light matter biologics to keep re-imprinting essence experiences and potentials that need to be set free. _**Aqua and Iris blues**_ give quality to truth and its focus as to why you created your universes. Its shades, hues, and tones assure that heart communication is always in instant flow; thereby assisting in inter-species communication networks The _**orange and peach**_ create qualities of celebration of light in all relationships, especially inter-stellar communication networks. _**Silvers and gold**_ can create exotic particles or; combine all qualities of magnificence and self-love into instant manifestations. The silver assures no energy will be trapped by the source that created it. A golden heart frequency manifests directly from heart awareness. _**Fuchsia magentas**_ create innovative versions of collaboration and unity that each is a unique Divine human and citizen of the new cosmic race!

**Hearts can then share** the self-realizations of their experiences and the conscious awareness of these new senses in their manifestations. Self-realizations and shared awareness replace any wounded thoughts, feelings, attitudes, or belief patterns from the past. So, this is new territory for marriage, relationships, and intimacy as shared soul essence in the education of the attributes of light. **Each soul knows they are first responsible to answer to their own energy and their own heart such that they never put self or other in a wounded, limited, or suffering relationship.** *They know the heart is in service to all life by fulfilling its unique myriad essence attributes and new sense qualities.* They let themselves love themselves and through masterful communication with the energy of the heart potentials in moment-to-moment creations, they know what really matters. This heart openness synchronizes all life! Unlimited Source energy reads and matches their ever-increasing heart capacity. A willing Heart_chooses and all life benefits. Allowing hearts support choices in equality is **faith and trust and** streams/transmits new consciousness to all life. This offers full opportunity and participation in the life that is lived in equal light relationships shared in new conscious essence attribute awareness. It also offers all relationship to join in the simplicity of heart's continued light fusions where **relationship always creates**, instead of separates and divides. This also eliminates the old Earth abusive or punitive communications or imposing systems that try to fix, heal, use control, or judge. These old Earth energy power systems are replaced by a multiplicity of diversity of organic creative choices to create new **innovative relationship styles and relationship light-fusion systems** for new lifestyles and species.

Notice how light fusion is fracturing space-time as all past-future time lines/existences heal and converge in new reality experiences offered by the light universes. Its language has already entered your **new careers and lifestyles** as in: designer imprint/replication technologies, plasma magnetics, Astro-sonics, interstellar commerce, intra-world or stellar inter-species sociocultural communication networks; cyber-sonic spaceport systems, or lifestyle mentoring and fulfillment centers.

# What did Cosmic Heart-Essence Change? 10/2021

Energy Communication Life Masters, Energy Potential Magi, Love and Life Fulfillment Masters, what has the light illumination of your Cosmic-angelic Heart Essence done? Has suffering, death, disease, and abuse enslavement, inside separation from soul, grieved enough of humanity's tears? Did global hearts change awareness; such that your transmission of enough quantum/meta-senses in the new human light form, can allow the entire cosmic reset another quantum leap in 2022, and even surprise love itself? **How many hearts opened self-love and self-acceptance of all life without judgment?**

Enter Aquarian Age. How much did Humanity change when their new viral consciousness activated their Core-Essence (crystal/ diamond-plasma) cell DNA? How much changed as global Essence Hearts grieved all their mis-qualified energies of abuse inside separation of self and all life; including being seduced by its fear, control, and mental programming; as if creation and energy itself could be forced or hijacked. Indeed, has humanity stopped abusing and suffering in love along with the **worlds within them**? Did they open their **mirror's awareness** to remember that they are first and ever Divine beings and can't lose the innate organic essence bonding code of their DNA stem cell's bio-love? Did they use their new 2020-viral consciousness to awaken their DNA-soul or use fear of it to cross-over? Or, is the new consciousness virus still lifting all humanity's shadows into the light for healing? In their isolation and loneliness, did humanity become renewed lovers of life's energy, committed to their own hearts as their true beloved, through this unique soul experience of growth? Or, is it all still too risky to lose material power over soul awareness, abdicating choice and staying in a conspiracy story of a war of souls? Are they still willing to trade their soul for disjointed technologies inferior to their light vessel's trans-sense capacity? Would their trauma trade their suffering for a glimpsing awareness that everything is their own energy, happening inside their own consciousness? Anything outside that could be accepted as movie participation in the illusion or theatre of another's soul's reality; or creation propaganda of broken distorted-conspiratorial life systems? How many life flames were extinguished by fear? Would they be banished from all the love they could hold, nevermore to be the tender caretakers of their own love, joy, passion-creations; by squandering their energy in judgments over experiences their soul's chose, in quest to master all life? Are their minds still torturing them or did they finally listen inside their hearts' core Essence Presence to the whispering in their ear; that it will caretake, nurture, and love them if they embody life's existence, with no need to validate or justify their worth, except to experience it all and self-love/self-accept without all their judgments?

Will they bastille the light revolution of change; where they will no longer give their love and worth away to the viral feeding energies of power agendas and controllers, who would barter their right to exist by selling them fear propaganda, and call it commerce. Has it has come as the need to have **a passport** to **validate** their equal and divine right to the unlimited resources of creation; offered by the freedom of the awareness of their own unique soul creations? Will they channel all this into creative solutions or still riot at what was withheld; and being deceived by the gaming systems they blame for falling into illusion? Who broke their essence bond to creation? How many hearts changed this and accepted renewed life? How much natural light pushed all this heart deception-exploration to the surface? More than any time in the history of your local universe; humanity continues sensing the: symbols, images, codes, videos, stories, or internal aspects of the multi-holographic layers of everything that they created. And now, their human soul's grief and anguish remain the underbelly of its violence, anger, blame, shame, hurt, abandonment, doubt, judgment, regret that were never resolved in fearing their soul's own energies and experiences? These were the unresolved trapped energies in your universe of spirit essence and angelic senses that later densified into negative human emotions. *Did humanity's love and light expose them? The opportunity to go inside, be still in their spirit's heart, connect with loved ones and be responsible for one's creations and energy; has offered incredible opportunity to realign Soul purpose across your Universe.* The limited human mind still clamors for the old safe predictable life, the old lifetime identities, past/future life times, wounded patterns, old spirit families, old limitations. The mind rages fear but projects a false safety to keep control of the human; for that is all it knows! Mind will: throw temper tantrums, anxiety attacks, fight with its own selves, act out addictions, or go insane; thinking it's keeping the brain and body safe from death? But mind cannot cure mind or body, only spirit heart essence can change cell matter. Indeed, meta-sense heart of the light body knows what it needs to self-sustain its own existence, as it is natural in the essence-DNA code to follow this. It was the past suffering of a distorted creational story where competition and separation became a suffering mind that doubted the heart of Creation's bond to Essence; rather than the gifts of receiving all that creation has to offer. Humanity had come to accept limitation of every form of fear as a prison of trapped distorted energies they would never escape and trapped in a body that lost its **soul's self-healing communication as a biological organism.** The essence-heart was never meant to be a symbiotic for alien energies who did not accept or understand that a new genetic cosmic race was the Creation code for this Universe. How many hearts in your universe changed this? But didn't rising transparent light allow your human eyes to see Jupiter and Saturn in your night sky in July?

**How many hearts** opened and allowed the DNA bio-cell transformation of soul's wounded core pattern to dissolve and re-imprint heart's new dormant genetic life codes? How many tears cleansed the wounded core pattern of: the right to exist without having to validate yourself, disappear, hide your light, prove worth-value, pretend; or empath for other's suffering in the constant dying of the old body, wounded lifetimes, fragmented DNA; unable to trust a heart that seemed always searching for fulfillment that was unattainable? Did the new **heart take off its mask** and awaken and say no more to anything outside its nature? How many still cry out to the heart of their spirit; "Please don't let this body-mind game and its nasty limitations blow me up, make me insane, or keep the polarity arguments torturing me like a robot, before I manifest heart's most fulfilling potentials! And maybe I might even meet my true self in a burst of light?" Did humanity bounce off enough polarities of: thoughts, feelings, attitudes, beliefs and choices, to finally collapse into their zero-point/core light? Did your world in all that viral time alone with family, with Self, without work, without distractions; face self without a mask? Did they reset their priorities? Did they realize that real abundance, real manifestation is self-value, self-acceptance, self-awareness, and self-trust of their spirit to steward their own love to fulfill their innate potentials?

Did humanity's soul indwell in a vessel of light they created for it; by mastering all human space-time experience in dark/light nights of their soul, which was recording every experience along the journey? And was there a hush all over the world where, humanity's quite spirits seemed so happy in the Presence of stillness; to reset their hearts journey and review all the growth, movies, collective chaos, and wisdom since it was separated from its soul and human aspects. And didn't Spirit even offer its Human-Soul a new heart biology of light to house its full Presence of light's quantum rainbow ray spectrum? Humanity had to experience a **soul-merge** of one or more aspects of self, even though they weren't sure who was sharing the body with them other than the mind. And now, they must consider life from having a soul or loved ones who did as they watched their loved ones cross over. Now what was life all about and was there some master code that bonded one to creation in some mysterious formula for life? Did everyone have it or was it just for secret governments, aliens and gods, or AI special humans? Who had the power to *the secret sauce of science, technology, or physics or even beyond all three*; that would give you all the treasures only stored in heaven? What did humanity discover? Is self-love too demanding? Was awareness in the heart enough to change the laws of science, physics, and technology **or** were there going to be decades more of slogging out realities; What did their hearts accept? Would they sense that their core essence light is naturally the tender care taker of their love, joy, and passion or continue to expend all your energy on judging Self and the Cosmos for the experiences they chose to explore the essence of themselves in living matter?

And what did you **Energy communication Life Masters transmit** to humanity through your streaming consciousness during the planetary reset? What did your hearts change? Even for many of you trusting your Spirit's Presence remained a stretch, as you fluctuated your multi-variant frequencies inside the physical and non-physical light spectrums. And just when you felt you fully integrated your light body, another aspect in the higher Old Earth realms of light who had never experienced density, needed to share its movies to be bio-merged.

So, juggling all these multiple-life aspects calling out for integration in your light body to dissolve polarities was consuming; and many of these aspects were never fully embodied before. And even now some master lights may still try to hang on to the old human form that the mind is terrified to allow to transform. Indeed, allow any distorted memory pattern in every cell as it cries and dissolves; so that in the new vessel, Spirit's love in the cell is sense-bonded in the DNA/RNA master code. Aren't your teachers of experience, fear and love, just energy seeking freedom back to essence? Then, **Heart awareness returns** as a new experience, a new potential, a new manifestation. Then, Heart awareness frequency, is/and equals, instant manifestation of heart's joy, fulfillment potential and creativity, as the natural bond to your own creation, as the Divine being you already are. Your consciousness messaged to humanity that; your Divine spirit indulges passion in enjoying being human while simultaneously birthing many new inner worlds of momentary experiences with all their various new qualities, growth stages and light particles. And, when that moment/creation is experienced, it dissolves back into free energy! Indeed, a moment is as a world born, experienced, enjoyed, transmitted throughout the cosmos, and released back into essence. A quantum awareness could be the sound of quantum kindness or the taste of light particles. Bonded in a composite true stable light quotient; or frequency of your own Soul and Spirit IAM, while embodied in a new vessel or quantum density, allows all potentials direct experience of their new essence flavors. Here, you know the result is that the heart's quantum essence-density can change anything is its own creation chamber, and enjoy blending attribute essences to create new homes, technology imprints, or organic matter. Your awakening has whispered that, "Whatever the heart is aware of in expression, joy of life, materiality, or creativity, **manifests directly out of Trust and communication of your own inner connection** and the inner bond to your Heart's excitement or inner chamber stillness."

This allows your spirit to reveal each upgrade step by step as the old vessel gradually re-particles into the new model light body.

We reiterate, that by embodying **the bio-instrument** that you have prepared for Spirit, thereby releasing the Old Earth Universe; you message a new stream of consciousness. There is no longer need for validation of your existence over and over in lost identity, purpose, or lost connection; for that would again, make you a victim of creation itself, and just be another distortion. That is why you allowed yourself to experience life review after life review, as you went back and forth to your version of inner New Earth and reviewed the movies of your life on your Old Earth simultaneously. *This tele-transport and light travel in frequencies opened your true composite essence vibration to remember your first light as natural Essence. You remember and did transmit to humanity; that you don't need to control, power up, or manipulate anyone or anything to allow yourself to exist and receive new life. And that life is You, because you have mastered all life. That's natural evolution as star seeds and midwives of worlds, seeking always to evolve creation codes to spawn and secure new life. This you illuminated and transmitted to the world and all its sciences, technologies, and energy physics for accelerated change.* Your energy is free and you're the everything of your own creation. In any master DNA life code, it is unnatural to steal energy or bargain your soul to access creational energy communication. It is also unnatural to carry the dense gravity of ancient ancestral DNA cellular grief from all prior generations. In fact, you light vessel's biosphere denies any alien or unnatural bio-signal, and is anti-viral, anti-time warp, and anti-cyber. That is because **gravity in your light** comes by allowing the Heart and all wounded feelings to arise in your awareness so they are forever: released, dissolved, re-molecule/d, re-essence/ed, re-imprinted, or changed by the natural heart code. This composite Heart with a bigger light capacity transmits your frequency, which will magnetize awareness across your world, such that that food can grow itself. The DNA can re-splice and grow new helixes. The rain-light can purify any toxins, and your vessels simply snack on the energy of pink turquoise joy. It makes a new space to be filled by heart awareness gifts and the potential of fulfillment in any moment Heart's light matter imprints a quantum sense. How about tasting pink chocolate covered compassion or tasty velvet pink sparkled sound? Can you jump realities in green-lavender magnetic space stairs; or imagine objects that float to you? How about the artistic-sense of climbing into your own painted vision of what your life can be, and there you are, inside that **reality in a virtual moment?** And indeed, it cancels all other realities. These new meta-essence qualities of quantum-density, allow the atom and the quark-particles to dance in new creative quantum delights. Your Divine spirit indulges passion in enjoying being human while simultaneously birthing many new inner worlds of momentary experiences with all their various new qualities, growth stages and light particles. And, when that moment/creation is experienced, it dissolves back into free energy! Indeed, a moment is as a world born, experienced, enjoyed, transmitted throughout the cosmos, and released back into essence. Now you have the imprint of Divine-Human, who can enjoy all new worlds of light, **with New Earth school as a spaceport base and use your own bio-light vessel for light transport.** Your **Creation-code ensures** a Divine-human unique soul will fulfill all its potentials. So did your heart awareness lift the quarantine off your creations, to knowing, that all will be healed as your compassion gene wipes the salt-crystal tears off humanity's eyes? Then they can see into their heart's light, sensing that the heart of creation lives inside all souls and its heart is the only way in or out in unrestricted, freedom.

**Energy Potential Masters**, in the New Earth Chronicles of the light body, you have finally arrived in your own true reality and in anew human form. It is measured by the alpha light-body constant or your unique Heart signature's creation code. This is your atom's soul mass to light ratio, measured parallel to your unique new spirit's light to mass quantum ratio. All you're Stories, all your conscious dreaming, or those you meet when you pretend to be asleep; or in stasis from this reality are in potential realizations of their experiences. They can all remain **versions of you** until you collect all your selves into one new Master Self. All your poems, your holography of self, your perceptions, and your awareness take on multi-self-realizations in the continuum. That remains until energy super-sense communication in your Divine-human knows, that nothing else matters, and that it's Spirit will never let it go without. Then whatever is in the heart awareness, IS! **How many Light beings have accepted such self-love? Life fulfillment Masters,** now you know <u>why</u> the new heart consciousness in the Divine-human light vessel bio-sphere can transcend all physics, science, and technology and become a new genetic cosmic race. It is because it has evolved a new consciousness in a new Human form, to embody soul-spirit and enjoy new meta-aware potentials in its very own energy and conscious creations, and self-love adventures. And that, is real success, real joy, real fulfillment; and real-life mastery. We ask again, how free and liquid do you transmit love masters; that all of the world can sense, be changed, and choose the social capital of joy's sharing in equality rather than the value of enslaved consumption. Did your heart transmit that if they would allow all the love they could hold; they can experience every possible potential their heart could create? Did you leave a living imprint that, the only thing that Creation really cares about is experiencing Essence Heart **with and through each soul's new mandala of life?** As we have asked many times before; have you liberated you're love so deeply that all your other new worlds appear and you disappear into the light, never seen by the human's naked eye or even the eye of even artificial intelligence? Did you illuminate your core light, your hearts, and your consciousness so deeply that global heart of humanity awakened to the miracle that they are **the secret sauce** for a new genetic/cosmic race? Their miracle-sauce, it that they are Divine-Humans who have soul grown a super-conscious, meta-sense, meta-aware Heart with elegant artistic quantum senses, which can sense any energy and change anything in a moment-potential of essence awareness! And, each soul must access this miracle within the self-acceptance /self-love within their own soul heart. A new heart awareness is a new experience, a new potential, a new manifestation, and anew moment. In your light vessels, your heart biospheres a moment is a world born, experienced, enjoyed, illumined throughout the cosmos, and released back into essence; such that you can even taste the secret sauce that no cosmic being ever has before. Indeed, this is what your **globe's cosmic heart has changed** and its catapulting quantum awareness is taking humanity there. So, is the **heart of humanity, though your self-loving illumination as living examples**, ready to trigger global/universal heart awareness so deeply, such that each Heart's bio-sphere becomes its own greenhouse, doctor, loving partner, technology, or new galaxy? What did your heart awareness change or manifest now? What light transport does your heart gate open to: Human, Divine, or walk in and out of both at the same moment; or go beyond, and become **its own inner-world biosphere**? Are you ready for all the **new living matter biologic potentiates in** your biosphere vessels as you explore how your new cosmic race and all your: crystal, plant, animal, and new genetic and hybrid species **live in the light**? What clothes, homes, lifestyles, foods, light vessels, technologies, relationships, families, and new interstellar life systems will you design? And how will you use your beautiful planet as a spaceport for such light vessel transport in your biospheres? How you live in the light vessel is indeed the next adventure.

# Accepting Self and All Life
## in Age of Light 1-2022

Energy Potential Heart-life Masters, in the eyes and the heart of Creation all experiences are equal. Accepting Self and all life, **is how life will be live**d, in the Aquarian age of light? Only a perception of judgement, that life is limited in any way, could cause an experience to be separate from self or any part of life. **What is life without Judgment**? It is the *All THAT IS*, which includes the *ALL THAT ISN'T*. **With judgment**, heart essence communication has experiences that limit or lock out natural manifestations. This is because judgment inhibits the free flow of the heart cell's natural atomic energy (of experiences of both positive and negative: thoughts, feelings, attitudes, and beliefs); to resolve polarity issues and experiences to balance and neutralize, any harm or heart vulnerability, to self or life. Therein, soul's growth of self-love/acceptance, has natural built-in boundaries and discernment of experiences. A Safe heart is innocent, vulnerable to accept more love, and lives in playful imagination and regenerative freedom. Hence, judgment inhibits choosing solutions for growth and locks out new experiences of life's fulfillment potentials. Judgment further misreads; that soul-spirit is separate from itself as a Divine Being having human soul adventures. Judgment locks out the **marriage** of the quantum non-physical worlds and the physical atomic worlds in the new cosmic race you have created where new multi-diverse bio-**helix imprints** grow and create meta-sense imprints.

To **Change wound, judgments, or separation from self-fears;** you remember that in the cosmos all time is now. There is no past/future. The judgment or wounds on your experiences, or the 'fall/separation from grace', caused a perceived time-space rip splitting now into multiple pasts and futures. So, when you want to change wound or judgment such as: blame, shame, rejection, doubt, abandonment, abuse, control, enslavement or any, (negative thought/feeling/attitude/or belief); you resurrect or revisit it from the meta-sense now-awareness of who you are; rather than the personage, form, or existence you were when your energy created the judgment. The higher vibration 'You' heals/integrates the lower vibration you. The **All seeing/knowing Eye** of the spirit heart can witness from a neutral now and re-image, re-essence, forgive, change, reimprint any experience, in the way you wanted the growth to be without judgment or wound; thereby, healing/changing its outcome, or dissolving it back into free essence energy. You have named it: neutralizing polarized energy, watching the video, getting out of the programmed matrix, or awakening the soul to trans-sense it from your its natural core light; which is your own pure essence heart chamber in the imagination of your own consciousness. This way you realize it your energy that created it and only 'You'; and nothing outside you can change it, but your heart essence, in moment-to-moment awareness. It was your story or book of life mastery, and your Heart can change it at any moment; thereby making room for another potential which comes from a higher vibration, rather than lower vibration that created the distortion or misperception.

*Now, what if you had <u>an experience with an awareness of judgment?</u> Then its underbelly of addictive mind-emotion, which recycles through the old earth energy collective unconscious of: anger, blame, shame, doubt, regret, suffering, death, and disease; would find its way back to its natural essence heart to be freed. Thereby, life's natural organic love can repurpose the soul moment to moment.* Your own energy communication would release any stuck/trapped energy by embracing everyone and everything as equal in experience? Could you then enjoy each experience in fulfilment, growth, teaching, and even in all quantum sense densities at once? Are love and fear just energies of experience that are ready to go beyond themselves into new essence awareness, which can sense any energy, and change anything in a moment's potential? Then, **Heart awareness returns as** a new experience, a new potential, a new manifestation. And, when that moment/creation is experienced and fulfilled, it dissolves back into free energy! Indeed, a moment is as a world born, experienced, enjoyed, transmitted throughout the cosmos, and released back into essence. Then, Heart Awareness frequency, is/and equals, instant manifestation of heart's joy, fulfillment potential and creativity as the natural bond to your own creation as the Divine being you already are. Indeed, the spirit child within you knows heart's\ imagination is everything. **In accepting Self and All life** in living potentials your meta-awareness moves polarity: fat/skinny, worthy/ unworthy, lack/ abundance, bad/good, light/dark moves into integrated blended senses and multi-diversity of experience, rather than recreating any wound or harm. Do you notice when **you communicate** in relationship with life with what you: eat, sleep, drink, take medicine, inhabit your vessel, relate to others or loved ones; or live with or without fulfillment or passion? Does your aware Self still attempt to accept/ recycle any experience where heart senses no excitement-potential of growing love in new essence/sense qualities of natural joy? Have you shifted to where awareness manifests as only experiences of full-on living? Is your creational heart embodied, in its new bio-sphere vessel, ready to exist in its very own freedom to grow love in endless communication and intimacies with 'All That Is'? Would that authenticate natural abundance, real joy, real fulfillment; and real-life mastery? Do you pulse life's breath, with the heart's magnetic pull, in acceptance flow open to new experience? Is there knowing the essence heart **can't be illusioned** by natural energies outside its essence communication with all life? Indeed, organic essence life always follows its natural life-code, and embraces soul's newly-deliciously grown, tasty essence senses; as your own **intimate relationship** with life. Is Exploring heart's passion potentials, in energy communication with self-loving acceptance; how you will live in the next cycle of light? Indeed, the **primordial crystallization process** is your natural soul's master code and monitors itself. Light children just imagine unrestricted Heart Senses as they communicate with all new life potential moments, inside natural meta-awareness, of blends of taste, touch, smell, sound, light and density. That's why children love their super-heroes. Animated biogenic cell, (crystal-diamond-plasma), essence love, now knows. It knows expanding love through direct experience, inside self-love/acceptance realization-awareness; follows its DNA Heart code as its only true guidance system for life's existence.

**Communication Heart-Fulfillment Life Masters**, you now can **shine your light and transmit the imagination and magic** of a planet-universe without Judgment, and its inherent suffering, thereby opening meta-sense awareness potentials for your new cosmic race. A new heart awareness is a new experience, a new potential, a new manifestation, and anew moment, such that, the **heart bio-sphere** is its own greenhouse, doctor, gene splicer, loving partner, technology, or new galaxy? Is your love so free and liquid that your food can grow itself, water cleanse itself, and you can regenerate your own DNA bio-helix imprints, to create without having any experience ever the same. How free and liquid is your love that it will grow more new adaptive light **helix imprints** with the built-in gene of compassion. Indeed, 'Awareness IS.'

What light transport does your heart gate open to: Human, Divine, or walk in and out of both at the same moment; **or go beyond**, and become its own **Source Biosphere**? Masters, ask your hearts again! As energy potential Masters, are you ready to fuse living matter biologic potentiates in your biosphere vessels; as you explore how your new cosmic race and all your: crystal, plant, animal, and hybrid species **live in the light**? What clothes, homes, lifestyles, foods, light vessels, technologies, relationships, families, and new interstellar life systems you will and have you already designed? And how will you use your beautiful planet as a spaceport for such light vessel transport?

# Awareness in Light Bio-genesis 2-2022

Energy Communication Masters, Energy Potential Magi, Love and Life fulfillment Masters, **how will you live in full spectrum light**? How will you live in the light and its new bio-vessel? Let's review the energy dynamics of the completion of the **primordial crystallization process of your soul's master code and the creator potentials you have brought forth now!** Indeed, you have mastered an **organic simulation of your Creation** where all possibilities and potentials of a unique soul can be experienced and grown. Quantum Attributes in Manifesting Awareness will lead your, (crystal-diamond-plasma), bio-genesis, in your light vessel stages. As, Energy Potential Masters, you're exploring how this elegant new heart consciousness in your Divine-Human light vessel will transcend all physics, science, and technology. Heart Essence has grown/evolved a new consciousness in a new Human form to embody soul-spirit to exist in its very own sense expressions and creations, of manifest love. And **such dominion** over the crystalline process, is real success, real joy, real fulfillment; and real-life mastery as a sovereign divine-human creator. Herein, the quest for the meaning of life changes. **To exist, experience, and embrace all life is its own meaning**. And, as you continue to transmit this to the world, then benevolent technology and light body sciences can be offered to all humanity so they can steward their planet and their own radiant health, love, abundance, and enjoy the genetic compassion of their own awakening master codes. *The <u>simple key</u> is a super-conscious or meta-sense Heart with elegant quantum senses, which can sense any energy and change anything in a moment of essence awareness! Then whatever is in the heart awareness, IS! Hence, a core light heart awareness is a new* *experience, a new potential, a new manifestation, and a new moment. Hearts very awareness is its own Source manifestation will eventually take humanity beyond physics, science, and technology by growing heart essence into your own bios-spheres, in full spectrum atomic-quantum blended light.* They will have liberated their love's awareness so deeply from flatline to flashpoint; such that all your other worlds appear. And you disappear into the light never seen by the human's naked eye or even the eye of artificial intelligence.

Hence, your planet is operating the **light transport base for the new cosmic race** light vessel prototype. Such is free energy in full out living, as essence heart communication within soul's living mastery of creation. So, is the heart of humanity **ready to manifest their new quantum attributes into awareness**, so freely; such that the heart of each soul has its own biosphere, and each soul-spirit has all life within it? It has its own oceans, greenhouses, physicist, doctor, loving partner, technology, or new star-sun universe; and anything you as a creator can imagine. How many are ready to switch reality back and forth between human and Divine, or walk in and out of both worlds at the same moment, or go beyond and accept their own biosphere as its **own Source spaceport and bioship**. Such is, direct access to 'All That Is.' No mathematics, log rhythms, technology, science, or physics can compute an everchanging Eternal Presence of the All That is. Consciousness is always in the Love and delight of its creations via their direct experience of ITS everchanging energy awareness of cosmic potentials at all levels of existence using free energy to communicate with all life. To grow and master your own unique soul has been Creation Heart's life code for each cosmic cycle until evolution is mastered.

However, your planet GAIA, operating as a **cosmic spaceport,** is still undergoing its re-education by the light or re-genesis, through the **repurpose of Essence** in every area of life, including the ecology of relationships with: people, places, events, systems; as well as all kingdoms of: animal, plant, crystal and all elements. This is the base for reformatting all your life systems in the light and allowing **your soul to be the living, playful, and creative art forms your essence was meant to be**. In the Old Earth Universe extreme polarity reduced relationships to programmed experience, forcing and mimicking creation through false essence attributes of mind, genetics, or technology, using illusions such as power, control, symbiotic feeding energy, addictive, or tyrant energies. It offered the illusion that love comes through a hypnotic-addictive mind, instead of the heart's natural organic **essence attributes** of natural grace, joy, compassion, innocence, and an excitement for creating. Creation story myth illusions have been a kind of spiritual propaganda of version judgments of truth using past-future: thoughts, feelings, attitudes, and belief amnesia games; of not being who you really are. Illusion-simulation experiences/existences, have offered 'Self 'against its own heart; accepting death, disease, or suffering to **replace your natural essence sense attributes**. But, extreme/wounded experiences, became dense reincarnated/recycled patterns or scared memory identities, which limited the essence communication bond to life, to your unique soul-spirit, and the Spirit of All of Creation; which_informed your universe of what its love was becoming through you. *You have also realized that you intended all illusions of experience to serve you as a contrast, instead of their judgment byproducts of extreme polarity*. However, your creative soul-arts have helped organic-essence memory expression of your love's aspects/existences. And their expressive attributes have helped heal past/future wounded time, and dense-solid-linear reality fractures. This allows all wisdom/experience free to flow into potential now awareness of all the different essence layers of the light spectrum, rainbow rays. or free energy bands that built your Creation.

So, as you descended into space time, they became dimensions that allowed you to express your **souls as living art forms**. **As Imagination Creators**, each ray or plasma **energy band** were really streaming consciousness, embodying descending life, as you called them the angelic realms of: elements, gods, and angels. You were all One, but unique in these blue prints as: gods, angels, and light ray bands. These were all your diverse creations that became images, symbols, blueprints, codes or imprints and later became as magnificent light languages of light, color, sound, soma/senses of all the diverse magnificent species: of crystals, animals, plants, elements and humans. All were carrying the **base DNA-life codes for a new crystalline cosmic race**. All life follows its creation codes. These were all available potentials inside DNA-essence attributes or codes of light, color, sound frequencies for each layer or dimension of soma-senses or streaming consciousness throughout the universe. **Like artists,** you created each layer in your universe in order **to** essence all aspect attributes in full ranges of expression in both quantum and atomic spheres/worlds. For example, the **color blue** has manifested itself and all its DNA ray attributes through you as its **life carrier** such as a: soul-seed light, as hurt or sadness, softness, a blue spruce tree, cold fusion particle light, free will, magnetic lazar vector, heart emergence; or even a cobalt blue crayon, or a sky. And yes, **you are these attributes and these attributes are you**. So, you as the creator of these attributes become the creation and the created simultaneously. Hence, as a life wave carrier you're enhancing your own soul and spirit's creations through evolving your natural organic senses to potential as manifestations. The soul thrives in growing awareness though direct experiences.

Each attribute such as: grace, love, joy, innocence, elegance, freedom, transfiguration, or infinite light are in their becoming and their own realization to inform the cosmos in its continual becoming and realization of 'ITSELF.' This allows your Eternal Self, a composite of your human-soul/spirit, to become its own manifestation and is the ultimate freedom and abundance. Hence **each attribute** is like a new flower, a new organism, a new helix which informs the cosmos what is becoming through its own creations and its growing essence attribute expressions. The creations of new Essence attributes are limitless because you have been growing them since you first incepted in this, bio-essence genetic Creation, 14 billion years ago. How many humans, soul, or spirit **lifetimes** have you created to enjoy yourselves in different aspects of these attributes to explore creation through all archıtypes of experience: gods, angels, humans, particles, animal, plant, and mineral species. Indeed, to live **every possible potential experience,** you as your own soul-creation can imagine. As, Master of Life's fulfillment Soul Heart knows it was indeed coded as the joy of Creation's every possible potential experience. In this new **enlightened consciousness**, the heart's light vessel no longer creates from polarity which deals with force, control, power, and opposites in conflict always seeking resolution. Your atoms and quantum particles have married and your meta light Quantum fields can go beyond time; and use stellar/cosmic light and sonar/soma senses and trillions of constantly self-realizing potentials, in bending and looping space and time to create new realities. These stream into your technology, physics and science upgrades, meditative imagery, symbols, codes, and sensory guidance throughout your moments of multi meta-sense-time shifting into new reality potentials. All torsion wave conjugations and particle interactions now enjoy this new quantum-density of the new Divine-human.

In the past many of you have been frustrated, because when others wouldn't receive your love or manifestation of these magnificent attributes; you thought you couldn't give them to yourselves; holding back your own light's love. This made you question your purpose and existence because you felt you couldn't fulfill your natural code's organic joy. This old misperception set the pace for looking outside heart for love instead of purposing your unique essence. Your consciousness created existing life for the purpose of growing essence potentials in a soul light vessel that could manifest Divine-Human meta-sense attributes, which was **your true purpose/code all along.** Your consciousness said to you that; "When you're Heart's Being-ness in its awareness manifests these new atomic-quantum attributes; then they are their **own unique encrypted soul code,** as and through you, at every level of existence. This Creation offered the All in All an opportunity to use your own soul's evolution to have LIFE self-realize what it is, how it manifests, and how it transmits/informs continued arising potentials. This forms a beautiful new Mandela of what joy and all other sense-attributes mean to each soul, so creation enjoys it awareness of what 'IT IS.' This allows each Source Creation to share essence and grow pools of multi-genetic diversity for all new life. Look at the many attributes of the marriage into quantum density,

A simple joy of **touching a snowflake**, emulates the journey of essence life descending from pure consciousness into an energy of existence. It's journey into plasma gases, radiation, light water, atmospheric condensation, water drops, is/was constantly changing itself to become the experience of its perfect code symmetry in creation. To melt upon the Earth, it had to become or interact with all the base fractal elements of the **primordial Heart's crystallization process of creation.**

Yes, everything in existence comes and goes from creation's consciousness and energy. Hence, the attributes of the atomic elements, have offered a base for the ability **to purpose heart's essence** and build your quantum instrument to **enjoy new**

**attributes** in direct manifestation of experiences of joy, creativity, expression, or materiality. These attributes grew from **becoming all aspects or archetypal experiences** offered in your bio-physical universe. This includes all the aspects of lifetimes and their attributes. You've have been negative aspects such as: blame, shame, judgment, pain, as well as elegant joy and playful selves. You have been the mothers, sons, daughters, lovers, children of creation. You have been doctors, priests, warriors, space commanders to unique your soul in life. Each experience offered new aspects with a multiplicity of attributes to master polarized time-space inside free will. You even grew a human form with: mental-emotional-physical-spiritual animations to accommodate any separation, distortion, or density from your essence's consciousness. You have played all roles with your soul and oversoul groups to master evolution of your soul's essence journey. And these new attributes of quantum density have arrived; such that whatever heart is aware of manifests! These dormant **new attributes** were re-birthed and through your 12 orbs, which replace the old Earth 12 chakra system, completing the Old Earth matrix universe. Your new 13th orb, as Sovereign Creator, inside a conscious heart, folds all the 12 orbs together into your new ascendant heart's (free energy plasma-particle light), or quantum-torus/biosphere vessel. It appears a new master-code for each soul. This is akin to mixing and matching the atomic rainbow light spectrum with the quantum spectrum rainbow to create new torsion rainbow spheres. Your unique new torus biosphere, can then form/experience, as many magnetic rainbow torus fields as you which, to explore all the new attributes and exotic particles for the New Earth universes and their new helix imprints. Your past, present, and futures are all going on now; thereby communicating/ merging/streaming into **One Life moments of particle matter interactive potentials.**

**Your** heart's master code now functions as stargate, magnetic transponder, regeneration chamber, and graviton field with new essence attributes in moment potentials; with its only purpose to fulfill. Each attribute informs the others and with any combinations of the <u>attributes and their meta-sense qualities</u> new life creates. Aspects and attributes of your atomic rainbow fields mixed with the quantum quark rainbow fields provide organic_combinations to produce new light life systems in **beautiful multi-tones, hues, and meta-senses.** This could be called **interstellar matter biologics or bio-genesis.** For <u>example,</u> all **shades** of <u>**yellow**</u> inside compassion can help build designer homes for those homeless from natural disasters. Shades of <u>**rose-pinks**</u> are inside tenderness or playfulness to help children feel loved and supported in their soul's gifts and their Earth walk. Attributes of <u>**blue**</u> inside the sense qualities of empowerment or emergence, can produce lazar blue inoculants and particle light inventions. All <u>**shades of white**</u> create qualities of transparency and new essence consciousness to reform your world's governing by its citizens. Its clarity shade offers interstellar communication systems. <u>**Greens**</u> are inside qualities of honesty and equal value to bring the justice of the soul for access to equitable self-sustaining life systems. It can also balance cyber-sonic frequencies. <u>**Lilac violets**</u> have all the qualities of release and unconditional freedom, so all life and species are valued in equity. It also serves in light matter biologics to keep re-imprinting essence experiences and potentials that need to be set free. <u>**Aqua and Iris blues**</u> give quality to truth and it's knowing as to why you created your universes. Its shades, hues, and tones assure that heart communication is always in instant flow; thereby assisting in inter-species communication networks The **orange and peach** create qualities of celebration of light in all relationships, especially inter-stellar communication networks. <u>Silvers and gold</u> can create exotic particles or; combine all qualities of magnificence and self-love into instant manifestations. The silver assures no energy will be trapped by the source that created it. A golden heart frequency manifests directly from heart awareness. **Fuchsia magentas** create innovative versions of collaboration and unity that each is a unique Divine human and citizen of the new cosmic race!

What one unique Master Heart-light fulfills, is meta-sensed, streamed, or illuminated to all hearts across your globe and cosmos. Self-realizations and shared awareness replace any wounded judgment patterns inside thoughts, feelings, attitudes, or belief patterns from the past experience. This offers aware choices for marriage, relationships, and intimacy as shared soul-essence in the education of the attributes of light. Each soul knows they are **first responsible to answer to their own energy and their own heart** such that they never put self or other in a wounded, limited, or suffering relationship. They know the heart is in service to all life by fulfilling its unique myriad essence attributes and new sense qualities. They let themselves **love themselves** and through masterful communication with the energy of the heart potentials in moment-to-moment creations, they know what really matters. There is no **artistic matter** in this world greater than such a radiant heart whose love can renew and re-perfect new life's dancing quantum symmetry, where **heart openness** serves as vulnerability protection, and synchronizes all life! Unlimited Source energy reads and matches each soul's ever-increasing heart capacity. A **willing self-loving and self-accepting Heart_chooses** and all life benefits. Allowing hearts support choices in equality is **faith and trust and** streams/transmits new consciousness to all life. This offers full opportunity-participation in the life that is lived; inside illuminating equal light relationships, shared in new conscious essence attribute awareness. It also offers all relationship to join in the simplicity of heart's continued light fusions and mirrored illuminations; where **relationship always creates**, instead of separates and divides. This also eliminates the old Earth abusive or punitive communications or imposing systems that try to fix, heal, use control, or judge. These old Earth energy power systems are replaced by a multiplicity of diversity of organic creative choices to create new **innovative relationship styles and relationship light-fusion systems** for new lifestyles and species.

Again, creation asks, are you ready to explore how your new cosmic race and all your: crystal, plant, animal, and hybrid species **live in the light**? What clothes, homes, lifestyles, foods, light vessels, technologies, relationships, families, and new interstellar life systems you will design? And how will you use your beautiful planet as a spaceport for such light vessel transport? Notice how light fusion is rapidly fracturing space-time, as all past-future time lines/existence, heal and converge in new reality experiences, sensed/offered by the light universes. Its language has already entered your **new careers and lifestyles** as in: designer imprint/replication technologies, plasma magnetics, Astro-sonics, interstellar commerce, intra-world or stellar inter-species sociocultural communication networks; cyber-sonic spaceport systems, or lifestyle mentoring and fulfillment centers. You are fully engaged in the next journey of your primordial crystallization process which serves now as a map for the whole cosmos and its new cosmic races; to exist, enjoy all experience, and embrace all new life essence you have been growing into the continuum of expansive love, since you left Creation!

# Awareness in Light Bio-genesis 2-2022

Energy Communication Masters, Energy Potential Magi, Love and Life fulfillment Masters, **how will you live in full spectrum light**? How will you live in the light and its new bio-vessel? Let's review the energy dynamics of the completion of the **primordial crystallization process of your soul's master code and the creator potentials you have brought forth now!** Indeed, you have mastered an **organic simulation of your Creation** where all possibilities and potentials of a unique soul can be experienced and grown.

**Quantum Attributes in Manifesting Awareness** will lead your, (crystal-diamond-plasma), bio-genesis, in your light vessel stages. As, Energy Potential Masters, you're exploring how this elegant new heart consciousness in your Divine-Human light vessel will transcend all physics, science, and technology. Heart Essence has grown/evolved a new consciousness in a new Human form to embody soul-spirit to exist in its very own sense expressions and creations, of manifest love. And **such dominion** over the crystalline process, is real success, real joy, real fulfillment; and real-life mastery as a sovereign divine-human creator. Herein, the quest for the meaning of life changes. **To exist, experience, and embrace all life is its own meaning**. And, as you continue to transmit this to the world, then benevolent technology and light body sciences can be offered to all humanity so they can steward their planet and their own radiant health, love, abundance, and enjoy the genetic compassion of their own awakening master codes.

The <u>simple key</u> is a super-conscious or meta-sense Heart with elegant quantum senses, which can sense any energy and change anything in a moment of essence awareness! Then whatever is in the heart awareness, IS! Hence, a core light heart awareness is a new experience, a new potential, a new manifestation, and a new moment. Hearts very awareness is its own Source manifestation will eventually take humanity beyond physics, science, and technology by growing heart essence into your own bios-spheres, in full spectrum atomic-quantum blended light. They will have liberated their love's awareness so deeply from flatline to flashpoint; such that all your other worlds appear. And you disappear into the light never seen by the human's naked eye or even the eye of artificial intelligence.

Hence, your planet is operating the **light transport base for the new cosmic race** light vessel prototype. Such is free energy in full out living, as essence heart communication within soul's living mastery of creation. So, is the heart of humanity **ready to manifest their new quantum attributes into awareness**, so freely; such that the heart of each soul has its own biosphere, and each soul-spirit has all life within it? It has its own oceans, greenhouses, physicist, doctor, loving partner, technology, or new star-sun universe; and anything you as a creator can imagine. How many are ready to switch reality back and forth between human and Divine, or walk in and out of both worlds at the same moment, or go beyond and accept their own biosphere as <u>its</u> **own Source spaceport and bioship**. Such is, direct access to 'All That Is.' No mathematics, log rhythms, technology, science, or physics can compute an everchanging Eternal Presence of the All That is. Consciousness is always in the Love and delight of its creations via their direct experience of ITS everchanging energy awareness of cosmic potentials at all levels of existence using free energy to communicate with all life. To grow and master your own unique soul has been Creation Heart's life code for each cosmic cycle until evolution is mastered.

However, your planet GAIA, operating as a **cosmic spaceport,** is still undergoing its re-education by the light or re-genesis, through the **repurpose of Essence** in every area of life, including the ecology of relationships with: people, places, events, systems; as well as all kingdoms of: animal, plant, crystal and all elements. This is the base for reformatting all your life systems in the light and allowing **your soul to be the living, playful, and creative art forms your essence was meant to be.** In the Old Earth Universe extreme polarity reduced relationships to programmed experience, forcing and mimicking creation through false essence attributes of mind, genetics, or technology, using illusions such as power, control, symbiotic feeding energy, addictive, or tyrant energies. It offered the illusion that love comes through a hypnotic-addictive mind, instead of the heart's natural organic **essence attributes** of natural grace, joy, compassion, innocence, and an excitement for creating. Creation story myth illusions have been a kind of spiritual propaganda of version judgments of truth using past-future: thoughts, feelings, attitudes, and belief amnesia games; of not being who you really are. Illusion-simulation experiences/existences, have offered 'Self 'against its own heart; accepting death, disease, or suffering to **replace your natural essence sense attributes.** But, extreme/wounded experiences, became dense reincarnated/recycled patterns or scared memory identities, which limited the essence communication bond to life, to your unique soul-spirit, and the Spirit of All of Creation; which informed your universe of what its love was becoming through you. You have also realized that you intended all illusions of experience to serve you as a contrast, instead of their judgment byproducts of extreme polarity. However, **your creative soul-arts** have helped organic-essence memory expression of your love's aspects/existences.

And their expressive attributes have helped heal past/future wounded time, and dense-solid-linear reality fractures. This allows all wisdom/experience free to flow into potential now awareness of all the different essence layers of the light spectrum, rainbow rays. or free energy bands that built your Creation.

So, as you descended into space time, they became dimensions that allowed you to express your **souls as living art forms**. **As Imagination Creators**, each ray or plasma **energy band** were really streaming consciousness, embodying descending life, as you called them the angelic realms of: elements, gods, and angels. You were all One, but unique in these blue prints as: gods, angels, and light ray bands. These were all your diverse creations that became images, symbols, blueprints, codes or imprints and later became as magnificent light languages of light, color, sound, soma/senses of all the diverse magnificent species: of crystals, animals, plants, elements and humans. All were carrying the **base DNA-life codes for a new crystalline cosmic race**. All life follows its creation codes. These were all available potentials inside DNA-essence attributes or codes of light, color, sound frequencies for each layer or dimension of soma-senses or streaming consciousness throughout the universe. **Like artists,** you created each layer in your universe in order **to** essence all aspect attributes in full ranges of expression in both quantum and atomic spheres/worlds. For example, the **color blue** has manifested itself and all its DNA ray attributes through you as its **life carrier** such as a: soul-seed light, as hurt or sadness, softness, a blue spruce tree, cold fusion particle light, free will, magnetic lazar vector, heart emergence; or even a cobalt blue crayon, or a sky. And yes, **you are these attributes and these attributes are you**. So, you as the creator of these attributes become the creation and the created simultaneously. Hence, as a life wave carrier you're enhancing your own soul and spirit's creations through evolving your natural organic senses to potential as manifestations. The soul thrives in growing awareness though direct experiences.

Each attribute such as: grace, love, joy, innocence, elegance, freedom, transfiguration, or infinite light are in their becoming and their own realization to inform the cosmos in its continual becoming and realization of 'ITSELF.' This allows your Eternal Self, a composite of your human-soul/spirit, to become its own manifestation and is the ultimate freedom and abundance. Hence **each attribute** is like a new flower, a new organism, a new helix which informs the cosmos what is becoming through its own creations and its growing essence attribute expressions. The creations of new Essence attributes are limitless because you have been growing them since you first incepted in this, bio-essence genetic Creation, 14 billion years ago. How many humans, soul, or spirit **lifetimes** have you created to enjoy yourselves in different aspects of these attributes to explore creation through **all architypes** of experience: gods, angels, humans, particles, animal, plant, and mineral species. Indeed, to live **every possible potential experience,** you as your own soul-creation can imagine. As, Master of Life's fulfillment Soul Heart knows it was indeed coded as the joy of Creation's every possible potential experience. In this new **enlightened consciousness,** the heart's light vessel no longer creates from polarity which deals with force, control, power, and opposites in conflict always seeking resolution. Your atoms and quantum particles have married and your meta light Quantum fields can go beyond time; and use stellar/cosmic light and sonar/soma senses and trillions of constantly self-realizing potentials, in bending and looping space and time to create new realities. These stream into your technology, physics and science upgrades, meditative imagery, symbols, codes, and sensory guidance throughout your moments of multi meta-sense-time shifting into new reality potentials. All torsion wave conjugations and particle interactions now enjoy this new quantum-density of the new Divine-human.

In the past many of you have been frustrated, because when others wouldn't receive your love or manifestation of these magnificent attributes; you thought you couldn't give them to yourselves; holding back your own light's love. This made you question your purpose and existence because you felt you couldn't fulfill your natural code's organic joy. This old misperception set the pace for looking outside heart for love instead of purposing your unique essence. Your consciousness created existing life for the purpose of growing essence potentials in a soul light vessel that could manifest Divine-Human meta-sense attributes, which was **your true purpose/code all along.** Your consciousness said to you that; "When you're Heart's Being-ness in its awareness manifests these new atomic-quantum attributes; then they are their **own unique encrypted soul code**, as and through you, at every level of existence. This Creation offered the All in All an opportunity to use your own soul's evolution to have LIFE self-realize what it is, how it manifests, and how it transmits/ informs continued arising potentials. This forms a beautiful new Mandela of what joy and all other sense-attributes mean to each soul, so creation enjoys it awareness of what 'IT IS.' This allows each Source Creation to share essence and grow pools of multi-genetic diversity for all new life. Look at the many attributes of the marriage into quantum density, a simple joy of **touching a snowflake**, emulates the journey of essence life descending from pure consciousness into an energy of existence. It's journey into plasma gases, radiation, light water, atmospheric condensation, water drops, is/was constantly changing itself to become the experience of its perfect code symmetry in creation. To melt upon the Earth, it had to become or interact with all the base fractal elements of the primordial Heart's crystallization process of creation. Yes, everything in existence comes and goes from creation's consciousness and energy. Hence, the attributes of the atomic elements, have offered a base for the ability **to purpose**

**heart's essence** and build your quantum instrument to enjoy **new attributes** in direct manifestation of experiences of joy, creativity, expression, or materiality. These attributes grew from **becoming all aspects or archetypal experiences** offered in your bio-physical universe. This includes all the aspects of lifetimes and their attributes. You've have been negative aspects such as: blame, shame, judgment, pain, as well as elegant joy and playful selves. You have been the mothers, sons, daughters, lovers, children of creation. You have been doctors, priests, warriors, space commanders to unique your soul in life. Each experience offered new aspects with a multiplicity of attributes to master polarized time-space inside free will. You even grew a human form with: mental-emotional-physical-spiritual animations to accommodate any separation, distortion, or density from your essence's consciousness. You have played all roles with your soul and oversoul groups to master evolution of your soul's essence journey. And these new attributes of quantum density have arrived; such that whatever heart is aware of manifests! These dormant **new attributes** were re-birthed and through your 12 orbs, which replace the old Earth 12 chakra system, completing the Old Earth matrix universe. Your new 13$^{th}$ orb, as Sovereign Creator, inside a conscious heart, folds all the 12 orbs together into your new ascendant heart's (free energy plasma-particle light), or quantum-torus/biosphere vessel. It appears a new master-code for each soul. This is akin to mixing and matching the atomic rainbow light spectrum with the quantum spectrum rainbow to create new torsion rainbow spheres. Your unique new torus biosphere, can then form/experience, as many magnetic rainbow torus fields as you which, to explore all the new attributes and exotic particles for the New Earth universes and their new helix imprints. Your past, present, and futures are all going on now; thereby communicating/ merging/streaming into **One Life moments of particle matter interactive potentials**.

**Your** heart's master code now functions as stargate, magnetic transponder, regeneration chamber, and graviton field with new essence attributes in moment potentials; with its only purpose to fulfill. Each attribute informs the others and with any combinations of the <u>attributes and their meta-sense qualities</u> new life creates. Aspects and attributes of your atomic rainbow fields mixed with the quantum quark rainbow fields provide organic_combinations to produce new light life systems in **beautiful multi-tones, hues, and meta-senses.** This could be called **interstellar matter biologics or bio-genesis.** For <u>example,</u> all **shades** of <u>**yellow**</u> inside compassion can help build designer homes for those homeless from natural disasters. Shades of **<u>rose-pinks</u>** are inside tenderness or playfulness to help children feel loved and supported in their soul's gifts and their Earth walk. Attributes of **<u>blue</u>** inside the sense qualities of empowerment or emergence, can produce lazar blue inoculants and particle light inventions. All **<u>shades of white</u>** create qualities of transparency and new essence consciousness to reform your world's governing by its citizens. Its clarity shade offers interstellar communication systems. **<u>Greens</u>** are inside qualities of honesty and equal value to bring the justice of the soul for access to equitable self-sustaining life systems. It can also balance cyber-sonic frequencies. **<u>Lilac violets</u>** have all the qualities of release and unconditional freedom, so all life and species are valued in equity. It also serves in light matter biologics to keep re-imprinting essence experiences and potentials that need to be set free. **<u>Aqua and Iris blues</u>** give quality to truth and it's knowing as to why you created your universes. Its shades, hues, and tones assure that heart communication is always in instant flow; thereby assisting in inter-species communication networks The **<u>orange and peach</u>** create qualities of celebration of light in all relationships, especially inter-stellar communication networks. <u>Silvers and gold</u> can create exotic particles or; combine all qualities of magnificence and self-love into instant manifestations. The silver assures no energy will be trapped by the source that created it. A golden heart frequency manifests directly from heart awareness. **Fuchsia magentas** create innovative versions of collaboration and unity that each is a unique Divine human and citizen of the new cosmic race!

What one unique Master Heart-light fulfills, is meta-sensed, streamed, or illuminated to all hearts across your globe and cosmos. Self-realizations and shared awareness replace any wounded judgment patterns inside thoughts, feelings, attitudes, or belief patterns from the past experience. This offers aware choices for marriage, relationships, and intimacy as shared soul-essence in the education of the attributes of light. Each soul knows they are **first responsible to answer to their own energy and their own heart** such that they never put self or other in a wounded, limited, or suffering relationship. They know the heart is in service to all life by fulfilling its unique myriad essence attributes and new sense qualities. They let themselves **love themselves** and through masterful communication with the energy of the heart potentials in moment-to-moment creations, they know what really matters. There is no artistic matter in this world greater than such a radiant heart whose love can renew and re-perfect new life's dancing quantum symmetry, where **heart openness** serves as vulnerability protection, and synchronizes all life! Unlimited Source energy reads and matches each soul's ever-increasing heart capacity. A **willing self-loving and self-accepting Heart_chooses** and all life benefits. Allowing hearts support choices in equality is **faith and trust and** streams/transmits new consciousness to all life. This offers full opportunity-participation in the life that is lived; inside illuminating equal light relationships, shared in new conscious essence attribute awareness. It also offers all relationship to join in the simplicity of heart's continued light fusions and mirrored illuminations; where **relationship always creates**, instead of separates and divides. This also eliminates the old Earth abusive or punitive communications or imposing systems that try to fix, heal, use control, or judge. These old Earth energy power systems are replaced by a multiplicity of diversity of organic creative choices to create new **innovative relationship styles and relationship light-fusion systems** for new lifestyles and species.

Again, creation asks, are you ready to explore how your new cosmic race and all your: crystal, plant, animal, and hybrid species **live in the light**? What clothes, homes, lifestyles, foods, light vessels, technologies, relationships, families, and new interstellar life systems you will design? And how will you use your beautiful planet as a spaceport for such light vessel transport? Notice how light fusion is rapidly fracturing space-time, as all past-future time lines/existence, heal and converge in new reality experiences, sensed/offered by the light universes. Its language has already entered your **new careers and lifestyles** as in: designer imprint/replication technologies, plasma magnetics, Astro-sonics, interstellar commerce, intra-world or stellar inter-species sociocultural communication networks; cyber-sonic spaceport systems, or lifestyle mentoring and fulfillment centers. You are fully engaged in the next journey of your primordial crystallization process which serves now as a map for the whole cosmos and its new cosmic races; to exist, enjoy all experience, and embrace all new life essence you have been growing into the continuum of expansive love, since you left Creation!

# The Shift of Light Body to Ascended/Stellar Form

Welcome Energy Communication Masters, Energy Potential Magi, Love and Life fulfillment Masters, and those awakening. We review the energy dynamics of your next **conscious energy-matter potential, which is the <u>shift of light body to an ascended form</u>**, or imprint vessel. In the new bio-organic species imprints as one new cosmic race for Humanity you are anchoring, a new light species genetic-code super-universe. Herein, you will be imprinting new conscious matter, of any life form, in a free energy vessel which then **grows into a <u>stellar form or biosphere of light within infinite unknowns</u>**.

Indeed, you have mastered your bio-genetic, experiment, and exploration of your Creation where all possibilities and potentials of a unique Divine-human soul can be experienced and grown in **biological love mastery**. In your newly born super universe/s, you live: beyond time, technology, physics, polarity, judgment, human mind-emotion/wounded limitations, science, and any distorted dinosaur systems, that interfered with natural evolution. You have mastered 3$^{rd}$ and 4$^{th}$ space-time density life. In the past-futures of Old Earth Universal DNA-codes your choices mastered singularity of: space-time polarity, limitation/distortion, fear-survival, illusion programming, death, and disease. Your own energy had the experience of being trapped or enslaved inside separation from direct essence communication with your own soul-spirit access, and the soul's recorder memory system.

Overall, the **ascended vessel** moves you beyond **light body which has been a transitional vehicle** to stabilize your regenerated core essence soul frequency imprint, until you could become your own free energy sovereign creator. In this consciousness into energy matter movement, the ascended form grows into your own conscious biosphere, using your heart's master code, to imprint any form of life in biological love. **This is the energy consciousness of the Divine Heart Essence growing the biology of Heart Love from the light body to the ascended form and beyond the beyond into the stellar biospheres of the infinite unknowns of light.** You can create worlds within worlds, reals, staircases, galaxies, or super-universes or any form of life you desire and when your fulfilled by the experience, return it to free energy. As, Gaia Earth moves fully into her light body, your planet also serves as a_spaceport_for light body transport, the ascended form, and beyond into your own stellar biosphere with the heart as its own mobile stargate_and center of gravity_

Therefore, New Earth remains a genetic universe and is being fully restored to genetic integrity. It's all part of disclosure and the truth of who you are as a species and what your IAM-DNA imprints now carry in your bio-physicals. Your fully conscious bio-physicals, along with the bio-soul of Earth are seeding all the new Quantum multi-helix imprints. These include the new Essence vessels and cosmic intelligences or quantum master codes to build new super conductive light systems as worlds created with dark matter. You are the Meta Universal School that you have all become. This is because full conscious embodiment has restored and grown soul's full Essence genetic integrity.

This was done as your Universal-Male RNA and Female DNA human emotions bio-merged with angelic senses that grew into new unique soul, meta-essence qualities, tones, and hues within the Divine Heart. Your genetic generations are moving into their soul's Essence helix imprint-bio vessel which operates as a quantum particle body with one Heart essence stem cell. All your sciences, physics, and technology will reflect this. Your neutrino embryo cell, which passes right through solid matter, allows you to change frequency and re-imprint your essence into any form, experience, or quality of expression you have yet to be; inside your own **stellar biospheres of embodied light.** This free energy vessel is your composite Divine-Human spirit embodied in the substance of Love, or light matter biologicals, where new awareness experience is a manifestation. As Gaia moves into light vessel she will also serves as a space-port for these essence biospheres which humanity still names as alien space-craft.

The **transitional Light Body:** resplices, awakens, integrates, and stabilizes its DNA codes and transcriptions exponentially until it becomes the new essence free energy vessel biosphere in all the New Earth Universes. The core essence of the light body rewires the brain until it is a data processor for the heart's light networks and it **has re-spliced its old ancestral DNA** into the upgraded and adapting master bio-life—code. Its **Essence DNA/helix imprinted heart cell** then acts as: transporter star gate, a magnetic imprinter, Source Code/r, centrifuge, quark stem cell particle and bio-ship for New Earth spirit matter, inside embodied love. **Light body switches** from the Old Earth matrix blueprints and mass programming to new light systems; which communicate and access the dormant quantum DNA blueprints and unique master code/imprints. This vessel in the multi-light universe is a blend of the physical and nonphysical into new conscious **superconductive light systems**. These bio-systems include new adaptive imprint Source code templates made of organic essence consciousness within the ALL That IS.

However, your **new species imprint** is a blend of Divine and human. It is a blend between seen and unseen worlds. It is a blend between the atom and newly born quantum light particles. It is a blend of a: crystal soul cell, a diamond spirit cell, a multi-plasma orb, and liquid light particle cell. It is a new heart stem cell that can regenerate, re-imprint, or repair your entire bio-organism right out of your own consciousness. It is a blend of Old Earth atomic and New Earths quark blueprints. It is a blend of Linear/singularity and multi/quantum applications of time and space. It is the prototype of a unique in-souled sovereign organic Essence Human-Master Creator fully embodied as atomic-quantum matter essence. Science has thus far called it dark energy-matter; that which makes new light universes!

**Light body Masters**, there can be some confusion about the instability of the soul's core light phasing in and out. This is so, because there is a major difference between the old limited energy and new energy creational aspects. **Old earth universe master codes included**: parallel universal lifetimes, life forms, bodies, patterns, blueprints, emotions, existences, or even senses you have lived. These participated in a distorted story of creation where the life code gene of compassion was suppressed and altered by giving in suffering; rather than receiving all the life that the cosmos wanted you to become and fulfill. This left the essence biology and its growing forms vulnerable to occupation or harm by others. Here, the energy of **service to self and service to other** went into competition and separation wound.

**The light body** allows you to gather all these experiences with all life and its species kingdoms into the distilled wisdom of your composite soul's heart's, mass to light ratio frequency code. It transforms all essence heart learned about its biological experiences into **new multi-trans sense attributes** of: tones, hues, shades, flavors, colors, tastes, sounds, in quantum density experiences;

replacing any wound from the trauma of human-soul experiences. You did this via your essence in: life reviews, images, symbols, dreams, or re-experiencing past-future memory outcomes to bring your soul's growth into self-realization, self-acceptance, and self-love. This Earth life offered every mirror of people, places, circumstances and events in the ancestral DNA human-soul/spirit families reflected/ mirrored in your daily life to heal any patterns of **wound judgment or core patterns of separation** from Essence SELF! Any excess trauma has grown the soul's, new DNA/ helix imprints biosphere; into a sovereign anti-pathogenic, anti-cyber, anti-time warp, **free energy agency**. Indeed, it was so that the soul could evolve its own unique love by mastering time-space and all atomic organic DNA-cell polarities; while deleting any distorted coding and **using biological love to return the essence** back to its natural state.

Have **gratitude** that each limited experience of these, human-soul-spirit aspects/lifetimes, who lived in dramatic experiences of shadow light reflections of subhuman acts and unnatural practices, and in addictive positive and negative suffering patterns; are the aspects of you, who grew your new essence potentials, now integrated into master heart code/imprint. As each of you Creators in the Old Earth Universe gather all your: suns, stars, aliens, fallen angels, galactic warriors, priests, starving humans, children, crystals, gnomes, ferries, grains of sand you created; your universe is set free to become new potentials or dissolve back into the all of consciousness. These new potentials carry the **essence of mastered biological love,** which originated from the original humanoid gene of compassion. This original life code kept giving and receiving equal, with harm to self or another impossible. It also was designed to protect against falling into competition, judgment, or enslavement. For, the 'ALL in ALL' learned from you; that dark and light were just reflections of each other in the matter-antimatter of the atom. This teaches that **nothing**

**can be created or destroyed,** just experienced and evolved via life's essence and returned to Eternity to begin again, or dissolve into free energy. All aspects of your life are loved for their service; because Creation loves all its creational aspects and evolves from their *storied experiences*. Technology and science are also stories in the study of this consciousness to try understand the cosmos and its conversions of energy and matter. And now humanity will use benevolent science, technology, and physics to go beyond space-time and trans-migrate to their New Earth homes. For, now all these old energy aspects have grown new qualities of Essence love to share with their star selves and new light neighbors. Maybe, pink jellyfish magenta peach tingles have become a new kind of compassion. Embodied/lived unique soul realities gives authentic Presence to the wisdom Master. Here, receiving your own creation is giving to the **cosmos multiplied in its own illumination**. You become what you create and you mind the business of your own creation. Receiving returns to giving inside illuminated reciprocity!

In your new soul's free energy master life code imprint, you are Source in form and the essence bio-matter imprint of your own creations. You were always responsible for your own energy and your own creations; although you participated in co-creation to master the art of creating existence. So, you either transformed your Old Earth universal aspects into this new bio-vessel composite imprint, or set them free while integrating their wisdom by reviewing their life review movies. This is what enlightenment or self-acceptance inside love offers you. Then, ascendant **Master Creational Self,** becomes and lives its own creations. They become **Source in form** by re-essence/ing or imprinting their master soul's new helix imprint wisdom codes into new energy awareness or expressions; that are playful and fulfilling trans-sensual expressions of any soul-heart fulfilling potential imaginable! Being Divine and human at the same time with interchangeable light spectrum rainbow-torus growth attributes inside the soul, allows quantum density to marry and birth new worlds of experience. For this is what trans-evolutionary creators love to do as the star seed and midwives of worlds. **Awareness truly moves back into imagination and receivership**. In receiving all that the cosmos has to offer you have grown compassion into biological love expressions that your soul essence has never know; and are now seeded/imprinted, in the Human-Angel-God Life Master prototype, for the entire cosmos.

So, **part of enlightenment**, is the <u>**constant self-realization**</u> from the newly born innocent spirit-heart, now animating the new species bio-vessel; while communicating through the wisdom of your composite Integrated spirit. Your Master Spirit is telling you, "Thankyou" for embodying the blueprint as a: plant, animal, male/female human, child, the wind, a stone, a reptile, an angel, a blueprint pattern, an energy essence, and creator. Such is the process of <u>**master imprinting.**</u> It offers <u>**all life new potentials.**</u> If you have been a physician, then Integrative Spirit Self can heal your cold. If you were a space commander, then you remember your brothers and sisters in the stars. In your divine memory as an angel, you know you can re-imprint any life form with your essence. If you were in Atlantis, then technology is an easy tool for you now, because you understand the soul is a crystal. If you studied in the mystery schools then all the ancient wisdom is available as to why you are here on Earth now. If you destroyed planets, then you understand the physics of atoms in creation, destruction, and regeneration. If you were a midwife crystal maiden; you birthed new DNA children that would ensure natural evolutionary life cycles. If you were an Elohim, then you know that you have charge over the elemental imprints or life codes of the forces of nature and the power of the elements; or climate changes. And when you remember that you have always been a spirit being, then access to the higher realms is always available through light travel. Hence, the import is that this composite self-realized or full embodied **integrated Divine-<u>Self</u> aspect**, is constantly drawing on the wisdom of all your experiences, to draw out new essence potentials in a continual integrative dialogue with your soul-spirit Presence of all new conscious life, as awareness then moves back into imagination. Awareness is a new experience. Whatever the new heart is aware of is! Awareness is a manifestation.

This is because the **<u>column of Light body</u>** moving through the spine becomes expanded into a giant full spectrum torus heart sphere or stellar biosphere of radiating light. The heart moved creating through focused intention of the atom into a quantum leap of self-realizing your everyone and everything, and 'ALL that IS,' such that, the cosmos reads and matches your heart's vibrational choices and they appear. It comes to you because your heart aware/s itself as everything! *Your IAM Presence light code awareness says," I accept My life and all life as it is now. Therefore, the mind is obedient and has adapted, and does not have to fight life, fight change, fight fears, or fight with the immune system.' And, the silver and gold essence cord are attached in the heart, not the body. Therefore: The light's embodied energy digests the food not the bowels or digestive system. And, its light breath your entire biosphere through the heart not the lungs. Indeed, your light code lives in its ascended instrument of the mastered self-love and regeneration of its existence.* This new energy communication is the voice of your **heart's knowing** that; being is now creating and every moment is consciousness meeting itself in new awareness, which is compassions journey into new essences of a new biology of Love! **Awareness is a new experience** because you know that <u>your world only exists now</u> inside your own biosphere as the Heart's instrument of consciousness. Soul-Spirit's Presence guides receiving in self-care, self-love, or passion-imagination with its fulfilling expressions or essence potentials. Receiving free flowing cosmic energy grounds and stabilizes you in your new free energy vessel, so your **heart becomes the center of gravity** and not Earth. Yes, then new **heart bio-sphere** opens its magnetic-gravity chamber to operate as its own star ship's warp drive. Then the imprint of form you are in is truly **an instrument** of your Divine consciousness. And because it is human and divine merged, it keeps you from recycling a pain body, past or future; or in conflict of service suffering, you have already lived.

However**, ascended Heart** enjoys all life physical and non-physical. It also enjoys chocolate cake, human skin, nature, relationships and all the intimate sensuality creative life has to offer moment to moment with a freelance energy script. Yes, both merges as human imprints as Divine. **Inclusive is <u>full access</u> to the quantum realities and higher realms in light travel or ascendant biospheres, which eventually comes into the awareness of the new light worlds and universes.** This True **vibrational Self** is now stable enough to embody you in your new vessel long enough to complete all possible potentials in the New Earth realms in the light body as you shift into the super-universe biospheres. You are the star, actor, and audience, of your own movies playing in the cosmic theatre. Your Composite Presence/Soul-Spirit will always guide you deeper and deeper into your own consciousness until you are **fulfilled to the brim** with every new essence potential you can imagine playing with, inside your own core essence biosphere. Indeed, your **innocent imagination** gets to play with its biological love and become all the new toys in creation. Imagination's internal voice of communication awareness is instrumental in reminding you that you are now here to enjoy the love you are. And, this gets **<u>transmitted into new</u>**: designer imprint replication technologies, plasma magnetics, Astro-sonics, inter-stellar technologies, innovations, starships, stellar commerce, and light body understandings of heart-essence relatedness.

This means human communication is now soul to soul and spirit essence to essence. This also serves the path of conscious technology for those behind you, since you prepared the way through **the path of embodied creation for finite and infinite heart's potentials** for all to follow. Technology serves as a bridge/tool to grow each unique soul, their own biospheres.

In sum, remember in the **super nova shifting** of New Earth, heart's biospheres; awareness returns as a new experience, a new potential, a new manifestation, and a new experience of biological love. And, when that moment/creation is experienced and fulfilled, it dissolves back into free energy! In your own biosphere a moment is as a world born, experienced, enjoyed, transmitted throughout the cosmos, and released back into essence. And, so it was when you first came forth from creation. Heart Awareness frequency, is/and equals, instant manifestation of heart's joy, fulfillment potential and creativity as the natural bond to your own creation as the Divine being you already are simply by allowing your **biosphere to shine its light illumination** wherever it travels in the cosmos! And within your biosphere will always be the imprint of the Divine human prototype. The key to the embodied realization of the primordial crystallization process, was always **true faith,** which is total self-acceptance and self-love without judgment separation from any part of creation. Now your natural soul's master code monitors itself. Again, as Gaia Earth moves her light body, your planet serves as a spaceport for light body transport. However, the ascended form or organic-biosphere grown in biological love is its own starship with the heart as its own mobile stargate.

**Remember, life is a heart fulfillment center** embodied Life-Masters. No matter what is going on in the worlds around you, no matter what goes on with your family, relationships, friends, career, and no matter what you see outside your own bio-sphere; your sovereign master creator has guaranteed that you will be fulfilled, as you allow its full embodied Presence to imagine with your new manifesting love. Your core Presence moves you moment to moment in your creational process constantly informing the cosmos of its own becoming. The Heart's new Cosmic race hears your elegant voice potentials inside inner master life-code communication from your Spirit Heart. It says, *'I love the way we live in our moment's essence heart core light chamber. I love the way we trans-sense, and are intimate with the images, potentials, visions, expressions of our next fulfillment, where we meet again and again in new awareness and in various new forms. I love the way we live moment to moment and enjoy ALL Life's imagination, by simply illuminating our light in pure joy!'* So, climb into your imagination and sense the light body moving into its ascended form and into its illuminating stellar biosphere to explore its potentials in new realizations and image awareness of love's play. It includes the base neutrino's oscillation, radiation, and illuminations experiencing new light matter particles, essences, and flavors of Source Creation in your cosmic age of light!

# Light Fusion Lifts Universal Mistrust - 4/2022 The New ascended Masters- through Maurene Watson

Quantum Masters, it is your **natural state** to trust Heart Presence to protect, provide, and guide life through you. However, the greatest universal secret on the planet, had been buried so deep in distorted ancestral DNA, that it has been as a masked veil blocking the IAM light of the human race for eons; and its natural primordial crystallization. This veil or mask, included both atomic anti-matter and quantum dark matter, codes and imprints; for all essence life's natural joyful expressions. It originated from Judgment/ misunderstanding that going into creation, from its core essence seed or spark flame, meant that the gods and angels lost or fractured their crystalline Essence. Herein, they perceived **being longer worthy to grow to love** themselves and all life, because they left home. However, now your aware, that it was just the All That Is Consciousness, creating the essence existence of your universe into its Essence sparks with all you gods and angels as co-creators to master creation. This ancient **mis-perception** of breaking the creation bond with heart essence made the need to validate and justify their IAM Presence and their right to exist a secret to protect the heart light from being vulnerable to self-hatred, self-sabotage, self-judgment, self-abandonment, or a venomous resentful anti-life heart now trapped in the energy of your local universe's great experiment. The Trapped energy deception echoed like a secret monster buried alive in a tomb of light and later a human coffin. It said, *"What if I deceive myself? What if I deceive another? What if I deceive Creator? What if they find out I can't grow love's gene of compassion.? Now, I have to hide my*

227

*light and validate/justify MY IAM to prove IAM worthy of love." Your aware now that your IAM needs no justification as it is The Presence of Creation and The All That IS. Trust in the IAM is an honest heart that says; "Divine Presence within, help me open my heart so I can feel what I'm not willing to feel. What an unconscious burden it has been in not being able to feel, sense, and essence love and be loved."*

Indeed, it's all in the willingness to open the heart and give any wounds to your Spirit Presence to set into free energy. This returned awareness is simply part of realization process going back to creation separation from core essence bond-light and its illusion of a broken/shattered crystal-diamond plasma source sun; rather than creation's Essence sparks creating the universe. It's truth surfaces, just as You as midwives of worlds, are again sparking the New Cosmic Age of the rebirth of a greater capacity of lighted essence-fusion love! Such distortion however, had masked Divine Human's beautiful changes your new consciousness has ushered in; such as benevolent soul-spirit technologies that humanity needs to assist their awakenings to remember and re-ignite their Essence. These **particle light fusion changes are arising** in the new light systems such as: DNA splicing, essence bio-matter imprinting, hovercraft and bio-plasma travel, magnetron manufacturing, hybrid AI-modulars, regeneration chambers, bio-imprinting, and other benevolent TEK that will bring unity and global equality to bear. Its language has already entered your **new careers and lifestyles** as in: designer imprint/replication technologies, plasma magnetics, Astro-sonics, interstellar commerce, intra-world or stellar inter-species sociocultural communication networks, interstellar contacts assistance and collaboration; as well as cyber-sonic spaceport systems, or lifestyle mentoring and fulfillment centers. This frees the soul essence to create, express, and play in its natural bio-aware consciousness.

This **is <u>allowing any mistrust and deception to be transformed</u>** by your light fusion, illumination, and transmission that, your individual soul's bio-ascension as your own unique creators is fully activated. As the pioneers of new Heart-DNA-essence consciousness, you are the template-imprints for the new quantum creation physics of new essence qualities of love within **ascended living.** All are experiencing cycles of transmutation symptoms on a daily basis that allow for immediate self-realized choices; as well as the vibrational stability of the Essence SOUL's particle fusion-light core. Changing DNA cell-essence consciousness of every life form in your world is the new norm, while harvesting new existences as Old Earth splits off into multiple bio-sphere realms, realities, or sub universes. Herein, bio-ascension is authenticated by the direct free energy expression of unique, sovereign bio cell-soul, (crystal-diamond-plasma imprint code), Essence DNA heart Presence, as the only Creator of its **own embodied reality.** Humanity has been dealing with their extreme polarity creation memories and timeline existences while you were master coding the new growth DNA-Essence heart imprint for the New Earth bio-fusion light network systems. Humanity had been oppressively lost in, a collective unconscious/unawaken mind-emotion state of extreme anger, within projections of their: thoughts, feelings, attitudes, beliefs, and actions. Their inner subconscious had felt abandoned, betrayed, rejected, and lost trust in their old dinosaur systems of: government, religion, money, and soul-less relationships in **<u>conditional potentials, not infinite potentials,</u>** of loving and goodness. They had been hypnotized to believe that the soul can be bought or sold by anyone or anything with or without responsibility or ramification of imbalanced energy exchange. Energy is free and must be balanced, and not trapped or enslaved. Humanity is releasing the old Earth DNA-fission memory that their existence is not deceptive betrayal and angry abandoned punishment by creation, because of the consciousness you are transmitting. The underbelly of abandonment is <u>unlimited fusion potential</u> of Life force energy.

The life force communicates energy freely, and is in an intimate relationship with each soul's inner Presence and the Eternal Presence of All That Is life and consciousness. **Creative free energy** is Life's Presence is in an energy relationship which is: alive, passionate, expressive, sensual and full of intimate-infinite essence potential communication. Consciousness is sense-essence energy expression including all the qualities grown by the soul's DNA-heart through human emotion, angelic soul senses, and meta-sense essence, spirit embodiment. Free Energy lives in the Presence of all Life's existence. Everything is energy, and energy is everywhere. Hence, the core spirit heart is the **soul's DNA essence energy imprint,** igniting new code potentials for sovereign mastery.

However, for humanity quantum-particle fusion-light is already here. They are overwhelmed with the choices of upgrading their bio-organic Heart Essence/meta-sense Human to enter the New Earths; or accepting synthetic-cyber, augmented, or virtual chip implants, for brain and body computer interfaces. This may include a myopic attempt in connecting all humanity as one Tek-mind and as hybrids or Techno-humans; rather than using Tek tools to assist activation of bio-conscious essence Homo-Sapien DNA-free energy stellar-sun vessel. Hence, time line memory warps and bio-particle accelerations are transmitting to soul-spirits that the vessel they occupy must have a base line DNA/compassion-essence cell crystalline structure. This includes continued acceleration, in order to naturally change bio-essence for their next existence to continue on into their soul's bio-enlightenment, or move into the ascended realms. If not, the soul can then enter a lighter, less dense polarity cycle due to their Old Earth contributions; and continue finishing soul and oversoul contracts, while serving the remaining old energy universes for their resolutions. Indeed, the growth potential is enormous as it effects healing all humanoid, soul, and spirit families as well as their original universes and their light multi-verse choice outcomes. As the average age on your planet is 25, your next light generations DNA is already activated to fulfill their DNA potentials for mass light migration, not extinction. All other Soul-Essence Star Families are waiting to interface with their New Earth families, for both interactive learning and assistance; and to experience the super-universe/s of light, since humanity is full of extraordinary entities from all over the cosmos here in your Creator School Universal experiment. There are master codes of every DNA species, as well as blended atomic-quantum color, sound, and soma light frequencies; that have never existed, making for extraordinary diversity within multiplicity. These are the **new infinite unknown essence potentials that** have growing since you incepted creation. The hybrid ET-Universe and all its simulations will be allowed to continue in parallel, until these creators and their manifestations have an opportunity to experience some light factor of essence matter in service to healing all life; rather than just forcing or trying to control evolution via simulated matter. Eventually, as New Earth Star goes into light fusion supernova, any life or dense matter not essence en-souled; will re-particle or be dissolved, and set into free energy.

Finally, humanity will expose the underbelly of their distorted/ re-spliced DNAs' anger and rage that fueled their old Earth cycles of: fear, hurt, rejection, depression, loss of power, addicted mind; along with the shame, guilt, and grief of holding back their soul light. They will take back their lives, their choices, their biological and bio-essence responses, their true nature to love, and **trust in the process of life again**. You are all too familiar with experiential memories of your creation myths, fallen-God and Angel stories; and their AI technologies, and hybrid aliens, in an attempt to replicate the human-soul essence for nefarious agendas. For, you have ascended these storied movies through your new master codes into meta-sense infinite DNA potentials. These old Earth Universe holograms may work for a while longer till Humanity discovers the true evolution and **disclosure of their species**, which truth is now clipping at a quantum pace. And what is being discovered is equally rich for new light fusion system applications. Humanity will finally feel or sense that they have been angry at everyone and everything, except what they are really angry at. That is mis-perceived primal DNA abandonment, mistrust, and deception by Creation and their own spirit for seemingly allowing them to be traumatized or abused by other creators or aliens, although they chose these experiences. This primordial anger at Source is recorded in notorious stories/movies of separation between the spirit and: its soul, and its human; and by the loss of divine communication, amnesia, and code distortion with the splitting or fragmentation of the crystallization code creation families and their parallel aspects. It's experience in a separation identity is recorded as fallen gods and angels who warred with ETs. However, in full potential this primal anger has transformed humanity into a new ascended species as a **new cosmic race and taught compassionate love** throughout the cosmos in ways never dreamed possible, even by the Ancients and Founders.

So, as humanity releases this ancient DNA/trauma and abuse; they are learning that a healthy distance of **boundaries** from family and others is not to be confused with abandonment. **Protection** is natural Heart communication with one's Divine Presence as its own energy and consciousness in its natural creation coding for life. They can now remember to **allow the abandonment its ascendant truth** so their soul-spirit can re-sense, re-essence, or re-imprint angry abandonment at cell level. Then, essence heart and only essence heart, can be changed into a new potential. This is the natural/organic right to your own soul sovereign: embody, emotions, choices, responses, passion and the **right to exist without justification**; or have your empathic sensitivities violated or overburdened. The imaginative Spirit-Essence children of Earth did nothing wrong and made no mistakes. Now humanity can stop abandoning their inner human child; because they remember their Divine Human communication inside the unconditional allowing of Father/Mother Creator parent within them, and their Creation. They can also allow their soul-spirit to embody in order to re-essence nurturing and re-parent themselves and each other. This transforms their emotional empathic confusion and over identification with those who control their biological responses, or bio-essence sensitivities, created as a tyrant-victim slave experience. This programing imprinted as an overwhelming distorted experience that felt like they had split off from their core-crystalline essence threatening the existence of their very soul. This overshadowed their true essence knowing; **that nothing exists or is real, except what is in one's crystal code-essence soul consciousness.**

As an energy communication Heart-Fulfillment Life Master, you are stewarding your universe's primordial-crystallization dynamic through its star-sun fusion; to **shine/illuminate your light and transmit** the imagination potentials of the light-fusion universes, without the judgment of mistrust and betrayal, and its inherent suffering. This opens meta-sense awareness potentials for your new cosmic race. Indeed, you're realizing that such heart-sphere awareness is a new world born, a new experience, a new potential, a new manifestation. It is its own gene splicer, food source, doctor, cryptocurrency, innovation, lover of life, ocean, galaxy, etc., because you incepted in the Imagination from All That Is.

Again and again, what heart gate light-fusion tele-transport does your divine-human open to? As you **travel in your heart awareness**

**you are there instantly.** Do you walk between the worlds or to go beyond, and become your own Heart Source Biosphere of star-sun fusion light? As energy potential Masters, you fuse living matter biologic potentiates in your biosphere vessels. And, as you explore these in your own consciousness and energy; you transmit and illuminate infinite free energy potentials for your new cosmic race and all your: crystal, plant, animal, and hybrid species moving into light? What clothes, homes, lifestyles, foods, light vessels, technologies, relationships, families, and new interstellar life systems have you transmitted for living in the light. And how will they use their beautiful planet as **a spaceport** for such *light vessel transport* that can go beyond all technology, science and physics? Remember what is natural. It is your natural state to Trust Heart Presence to protect, provide, and guide life! Whatever is in your heart awareness IS!

# Metaphysics of Love- The New Ascended Masters- Maurene Watson 5-2022

Welcome, Masters of Metaphysics and Bio-Essence Love, Light Energy Communication Masters, Energy Potential Magi, Love and Life fulfillment Masters, and those Light Beings awakening! You are now illuminating, transmitting, and streaming forth light fusion in a new standard of consciousness for a new cosmic race; through your free energy mastery of **the biology and bio-essence of love and its meta-physics.** Divine Self moves life's heart's bio-essence from the human to the soul's light body transition into the spirit's ascended form, and into the stellar/star-sun biosphere, which is a metaphysical or meta-sense form. This stellar star-sun biosphere is an illuminating Heart Sphere of light, which allows you to explore the new infinite unknowns of light fusion, to be lived as stellar beings of light; where you can light-travel, imprint-manifest, and create in your heart awareness instantly. This includes access to the higher realms, sub universes, windows, staircases of light, and super universes beyond the beyond into the infinite unknown of pure potentials. This light fusion as a divine human includes energy dynamics of your own unique consciousness. It is the shift of light body to an ascended bio/imprint or form-vessel, and beyond; into its growing stellar biosphere. Here in, you explore your own unique metaphysics and bio-essence meta-sense Love in infinite unknowns and as part of the new genetic/cosmic race. This is only/always within your own conscious awareness and energy or **illuminating Heart sphere of light.** Herein, light[fusion mastery over divine-human biology DNA/cell love, allowed the soul's bio-essence to experience and grow all life forms through self-love, self-acceptance, and self-awareness into a new standard or genetic ethic of love.

The gene of compassion has built in **genetic integrity** in the DNA-master soul code Essence. It has monitored the growth of the soul so it could be grown, rewired, re-spliced, re-essence/ed, restored, to master bio-essence cell love in free energy. This homeostatic dynamic returns any distorted essence back to its natural state; for its journey into the new light fusion realms, worlds, and universes. Therein, any alien: distorted, unnatural, or trapped energies could resplice, regenerate, and adapt all species life codes for the new cosmic species of light fusion.

Indeed, your universe's experiment, exploration, discovery, and next journey seeded as one new cosmic race, throughout the cosmos, came from mastery over the bio-essence of grown love. As these new galaxies, worlds, and light universes appear in your awareness, you will realize they are inside your own illuminating star biospheres as your own infinite unknown potentials to experience super-universal light fulfillment; which includes the ascended Divine heart-essence-human prototype you mastered. Your awareness is a base standard of soul heart **essence mastery in self-acceptance, self-love, and self-realization,** allowing Essence Heart to change atomic-quantum matter in any form of life via light fusion. So, why is any soul body, form, cell, imprint created so **important?** This is so, because the forms, codes, blueprints of creation, when: separated, fissional, or fragmented by their creators; allowed creational forms to become a battleground for distorted power over the essence master codes **to control existence and all the information within it.** It created an illusion that the Presence of Life, or all Creation, could be controlled, forced, or hijacked; against its DNA-code; to prevent Soul's free energy willingness, (free will), to evolve. However, the new light fusion's vessel standard bio-essence of love with its fully grown metaphysics; or meta-sense essence attributes and gifts of creation, **restores genetic integrity.** It also prevents any soul-essence form, body, imprint, or bio-organism from ever again being occupied, enslaved, programmed or controlled by an alien or distorted, illusions, or foreign energy outside one's soul consciousness and energy! **Heart Light's fusion illumination is its own protection.** This allows the stellar biosphere to create any new life form, universe, imprint, and experience its fulfilment, joy, love and play; and then release it back into free energy when the heart essence is full. Thereby, it remains a new potential transmitted, shared, and imprinted as a gift to The All of life and the Cosmos! Creation's infinite/unlimited imagination, inspiration, creativity, and joyful play via the heart reveals the difference between Old Earth-limited creation-communication energy and New Earth-unlimited creation-communication energy. Light body or **ascendant inspiration** comes from an essence heart standard of love. It's Isness can be quiet in creative stillness or regeneration. It can also be in its active energy play. **Life cares for life.** This Essence Heart, is just 'being' in creation, and inspiring and receiving its inner light fusion love illumination. It registers no heart posturing, mimicry, or imposing/harmful energies on anything or anyone. Spirit heart inspiration then, can be triggered by: imagination, an experience with your Divine Presence, communication with a child; by joy of a walk, a blue sky, the smell of a rose or perhaps a light travel experience communicating with the higher cosmic realms. Inspiration has a meta-sense awareness quality always arising naturally from the deepest consciousness of your essence-heart. Here no passport or justification of existence is required because the human soul and spirit wisdom experiences have been absorbed back into the light fusion embody to be used in new applications from your growing master code imprints.

Your wisdom **and embody of lifeworks** then, are about potentials in creative manifestation, creative abundance, and creative play. This invokes the 'Spirit Child' within the all of life with creative new careers within creative life fulfillment, and creative life relationships. You don't have to work at it. It is you, and comes to you, from your natural energy ethic of Essence heart's mastery over love in each DNA stem cell; as the maturing of bio-sense-essence love. Such energy awareness honors and respects your own feelings, senses, and essence in self and in others as they arise. This means allowing the human emotions, the angelic soul senses, and the childlike spirit essence that is the divine human to arise and guide. This is the natural Self, with no chasing the next or justifying; but connecting deep within the moments in the heart that guide your light fusion's illumination. Then **life builds itself for you,** and serves as an authentic living illuminating example, to all others in their life's awakenings and potentials. Here, there are no posturing or fake hearts; or controlling and judging in self/other, by not allowing or honoring the momentary old feelings of depression, aloneness, hurt, rejection, or denser feelings that mastered the beauty of having been a human; and taught the spirit how to love and grow the gene of compassion into the bio-essence organic of love.

**Most Old Earth human body imagination**, inspiration, and creativity came from human limitations of: mind-fear about safety existence, competition, control, feeding energies of external power agendas to compete; or be accepted by a mass programmed and limiting unconscious standards of success. The mind could rarely be quiet in the heart enough to feel that just being in creation's existence and accepting all life has to offer without judgment, was the true secret star to fulfillment. The **justification of ones right to exist** constantly trapped free energy of the soul. You are now well aware of the gravity entrapment drag of negative thoughts feelings, attitudes, and beliefs that needed the rebirth of awareness to integrate and re-embody all your human, soul, and spirit aspects that heart experienced; to feel self-love's fulfillment. Here in, you have **light fused, rebirthed, resurrected, and grown** the new heart essence species with a, regenerative divine-human/DNA code. Thereby, soul's core light vessel has deleted any **fission or fractured light**, that caused ancestral distorted alien or DNA that: would continue to impose an animal, violent, or of sub-human nature on the newly birthed **humanity and all life's species arising in fusion crystal-diamond-star-sun light**. And, any shadow, disgruntled, or imposter hearts that intend false power or harm are: exposed for their responsibility to heal, screened out by their own energies, and returned to the source that created them, or superseded by light fusion networks.

Therein, the standard of consciousness for discernment of love's light fusion and illumination has a built-in higher standard of how a master uses their own consciousness and energy. The Human-Master Heart quite naturally does not yield to old energies of psychology wound, mental analysis, an imposing or imposter heart, or interference on another soul's journey. They do not infringe, judge, heal, fix, or impose their energies in any way unless asked **to illuminate potentials for another's light** in another's awakening; therein triggering their soul heart's inner light fusion communication. However, they may collaborate in sharing new conscious transmissions or opportunities with other light networks and cosmic beings and universes who are aligned with and part of Humanity's light ascension. Overall, the use of their human-Master consciousness and energies are free to open up their own stellar star-sun corridors, windows, staircases, light universes, and infinite unknown of potential that are of greater service to all. This is so because each master-heart consciousness first experiences these visions, innovations, transmission assistance, and potentials; and meta-sense realizations within themselves and their own bio-essence of love first. Then and only then, are they authentic in their master soul's light illumination across worlds and the cosmos. Again, it's master heart comes from their radiation of growing light. It **magnetizes and amplifies the soul light** of those ready to receive or be triggered by such ever expanding light, experiential love, and triggering of the innate bio-essence natural gifts. These new infinite unknown potentials in the master code imprints of Divine-soul's bio-embodied potential are then available to all humanity and the All That Is!

A Love-Life Heart Master's continued growth in their heart's light fusion potentials and illumination becomes their greatest fulfillment and gift to all humanity; replacing the old way of suffering service work, energy, holding, or empathic enabling of another's energy responsibility. *This assists the New Earth-Star Gaia to fulfill its role as a cosmic spaceport for both the light body, the new ascended form, and the stellar/star-sun biosphere of sovereign light fusion illumination. It also assists each master to continue to fulfill its massive potentials on the New Earth and the new galaxies and universes that will be appearing in the super universes of light; which go beyond ascension into each soul's infinite unknown of potentials.* This is because light fusion mastery over the human biology and bio-soul-essence of love; allows the heart essence to operate its imprinted heart cell as a: transporter star gate, a magnetic imprinter, Source Code/r, centrifuge, quark stem cell particle and bio-ship for New Earth spirit matter, inside embodied love? Remember, **life is a heart fulfillment center.**

So, as, Gaia Earth moves fully into her light fusion Star-Sun vessel, your planet also serves as a cosmic spaceport for light body transport, the ascended form, and beyond into your own stellar biospheres with the heart as its own mobile stargate and center of gravity. Your interstellar space telescopes and craft will soon discover this, as well as; the supporting life from all your interstellar neighbors and all the new cosmic race universes awaiting your next journeys discovery and exploration into the cosmic age of light. In sum, we remind again and again, that in the **super nova star-sun shifting** of New Earth, **Heart's biosphere's awareness returns** as a: new experience, a new potential, a new manifestation, and a new experience of bio-essence love. And, when that moment/creation is experienced and fulfilled, it dissolves back into free energy! In your own biosphere a moment is as a world born, experienced, enjoyed, transmitted throughout the cosmos, and released back into essence. And, so it was when you first came forth from creation. Heart awareness frequency, is/and equals, instant manifestation of heart's joy, fulfillment potential and creativity as the natural bond to your own creation as the Divine being you already are simply by allowing your stellar biosphere to shine its light illumination wherever it heart awareness travels, creates, or manifests its soul imprints in the cosmos.

*Light fused creation, not fission creation, inside your continued conscious heart's self-realization, will then explore how your new cosmic race and all your: crystal, plant, animal, and hybrid species* <u>*live in the light*</u>*? What clothes, homes, lifestyles, foods, light vessels, technologies, relationships, families, and new interstellar life systems will you design on your planet and with your stellar neighbors. You can avail your beautiful planet as a spaceport for such light vessel transport; where you can access travel in your heart awareness instantly. Notice, how atomic-quantum particle light fusion, is rapidly fusing space-time, and all converging past-future time lines/existences to integrate and heal, and allow planetary and comic migration each soul light's fulfillment. Its language is transmitted in your* <u>*new careers and lifestyles*</u> *as in: designer imprint/replication technologies, plasma magnetics, Astro-sonics, interstellar commerce, intra-world or stellar inter-species sociocultural communication networks, interstellar contacts assistance and collaboration; as well as cyber-sonic spaceport systems, or lifestyle mentoring and fulfillment centers. Enjoy your own unique metaphysics and bio-essence meta-sense Love, where Heart awareness Is!*

# A New Earth Awakening The New Ascended Masters- Maurene Watson - 6/2022

I Presence the sun's blinking eyebrows rising on the New Earth.

I do not know who I was anymore, without an identity, without the need to sleep.

Are all my aspects many? Are they one? What has my Essence done?

From where and whose dream do I now come?

But I'm sure my soul and spirit's whisper have just begun~ again!

The journey has cast all the shadows aside.

Veils were lifted from my heart to other hearts.

Illusions were pondered and thrown to dragon's wind.


Is the emptiness I feel an opening in the belly of creation?

Never to agitate my human's prided ribs,

Never to push my spirit into hiding or rolling blackout futures.

Each day I merge and meet a new part of self I've known before.

But now they are closer and some are here; and some together in council.

We all move in and out of essence together.

She sighs, He sighs, and We bend light against the sky,

Hoping to merge, to sense, to titillate something new, not lived before,

A breath more passionate, more alive, freshly spawned:

Trying out new pieces and species of each other.


The parts that know to love and merge and blend again and again,

How do we sense selves as one in this new world?

Without the loss of self and without all those empty minds chattering noise,

We can hear Earth singing songs in her Heart chamber again;

And, only the soft candied taste of bubbling sound on heart remains.

Childlike, we are rubbing wild daisies against our noses,

Mud on our bellies; inscribing tattoos with blueberries on skin,

Left by the blue raven's gazing protective image, reflected in the new moon.

For already, the day has gone into timelessness, and everything comes alive to dance.

Everything is alive! And oh, the echo tones and touches the skin of uniqueness now!

# Multi-Matter is Alive in True Abundance 7-2022 The New Ascended Masters- Maurene Watson

*Q: How does light body create an abundant life in the Multi-D Master and change Empathic perfection stress?*

Multi-Light Potential Masters, Earth-Gaia is moving into her new energy vessel or light body with you, as Essence light creates new light-fusion matter for all life. New Earth Gaia has hosted all your soul Essence light beings as the last of the old Earth Human DNA races and the first of the New Earth human DNA races. Let's, for a moment, review her theatrical role in the cosmos. She has used her vessel as a grand host for the entire universe and the cosmos. She has been the genetic experiment to birth a new specie, using the gene of compassion to grow creation's love and new light matter for the new light multi-verses. She has offered refugees of the cosmos a home. Gaia had even allowed under free will, the ETS from the past ancient universes to be hosted here, while you human angels seeded the new species DNA here in this Earth Universe for all the future now's. She has been a recycling bin for soul's who have been unable to master the reincarnation cause-effect energy wheel in their own blueprints. She has acted as a comic sanatorium for cosmic refuges, lost, or abused races. Gaia has also been a special school for soul rehabilitation. She has housed the Ancients of Days, now, (The Eternal of Days), who have been in charge of the elder ancestors of evolutionary universes. She has hosted renegade ETs who manipulated the fallen angels into being their police force to facilitate their own genetic hybrid programs to replace or breed humans. These angels are learning well to never abandon their own experience. Many races of renegade ETs have tried to clone or genocide the Gaia goddess DNA into a false Father Universe for their own control agendas for millennia; since their DNA only transcribes a right to rule consciousness.

New Earth Gaia has been a school for any universe that has ever been created to understand and create dense matter within free will choice. She has sponsored co-creators to study creation and become their own newly born creator gods. She has allowed them to master all 9 electrums of creation in the atom as every possible: thought, feeling, attitude, belief, commitment, choice, spoken word, or action availed to the soul essence-expression of consciousness in existence. She has been a school for the mastery of singularity physics of time, space, and atomic matter. She has also been a training ground for creations' quantum essences to understand how to evolve a soul as its own sovereign expression and master its own molecular bio-essence animation. Her opera of stories is renown throughout the Meta verse. **You will continue to be amazed by the truth of who you are in the universes versus all the programmed stories you have been fed by forces who did not understand the nature of love.** Already astronomers have pictures that confirm your particle gem bodies are star seeds for trillions of new universes, galaxies, and star systems visible to your telescopes.

However, humanity has allowed itself to be deceived about the true cosmic history of its universe and who you are as a species. Give humanity the perfected memory of the quantum love codes they need to find brilliant solutions that only lighted love can offer. This is possible because your Light vessels have opened multi-quantum senses, multi-time, and multi-potentials in an updated master code bio-cell available to all. As you open your light vessels you transmit all the various ways that many of you and others will or have come into enlightenment and reveal the truth of your cosmic history. New light systems are revealed in the new studies of: art, music, ritual, myth, science, technology, astrology, physics medicine, and interstellar worlds and systems. This comes with the realizations that these disciplines have been brought here by all the civilizations of your solar systems throughout your universe. It's time to integrate the wisdom of the ages with your new visions. These **alternate versions of reality** will provide the highest biological potentials available to every species on your planet. All these visions of how to change matter, and allow it to come alive as free energy are needed, as the planet is allowed to regenerate health and beauty through its own core light body, just as you are as a new cosmic species.

But, in this now moment, you are moving with your visions in/out of your **heart's quantum still points** and light's new particles of creativity, as you respond to all the new bio-light networks or quantum fields around New Earth Universes. Quantum still points allow your newly born vessels light particle recalibration, star-sun gate alignments, heart chamber creation, and cosmic transmissions to imprint new potentials. Even the old Matrix brain's circuit breakers are worn out, because your light sensors keep short circuiting the electro energy polarity fields that stop the flow of free energy creations. Your new divine emotions or trans-sensual realizations of who you really are and what matter is; are evolving into new light applications already! This has been very confusing as **the multi-dimensional light body** becomes standard. Full conscious angelic senses have grown throughout evolution into an elegant vessel of light. Quantum senses or Divine emotions descended into soul emotions and soul emotions grew into human feelings. In these trans-senses you must get used to feeling all multiple choices at once. This can create temporary disassociation or confusion in all experiences and expressions, until the heart magnetizes; or till all choices settle into one moment of stillness or highest choice point in the heart. You have not felt all dimensional realities at once since you were in the angelic realms or what was called the 5th Density quasi-physical soul realms.

Hence, mention must be made here, of an old wounded pattern to be aware of and release with the new Multi-D Masters who are coming on line with their bio-light vessel channeling systems from the light multi-verses. They naturally experience excess **empathic distress and human perfection syndrome.** Again, this means, multi-masters feel all dimensions at once and choices can be confusing till clarity drops in heart to allow the perfect manifest for the now moment. Their floating potentials must be, trans-sensed or re-essence/ed in the heart, to transpose any polarity transmissions. If the master feels through the human alone, then they might be concerned that matrix abuse might recycle. If they allow feeling only through the Divine, they can only relate to those in their angelic circle of light to avoid old wounded cell memories, or alien abuse memory. Or, they hold back love perceiving over-sensitivity; that they can only tolerate soul to soul connections, in order to protect their angelic senses or consciousness. The only remedy to end either/or separation, is the merge of divine human light infusion.

**The old pattern** simply comes from trying to compensate for the human's ancient fears of abuse, being hurt, not feeling safe to trust experiences, or the human mind's compulsive need to be perfect beyond detail, to avoid bio-death and neural assault. It is also their angelic senses falsely perceiving the need to protect their consciousness and trying to master dense human feelings. Remember, the old human mind is a data processor; addicted to power and control; and can't feel, only mimic emotions, let alone sense heart. One can't enlighten through mind. This pattern in **Old Earth personal relationships** was to get your partner or others to feel the way you wanted them to, (meaning in their angelic senses), and see the universe the way you saw it, (as only light or an angle does), in order to be safe enough to have them love you. If they didn't then the cellular anger would arise to force love upon them, weather they wanted it or not. This old human mind addiction would mean you were going to love them, so they might love themselves, so it would be safe for them to love you. This created dependency, addiction, and fear in all relationships.

This implies a degree of **perfection stress** the human is not capable of, nor designed for. However, emerging Light body masters always attract their own circle of light and partners who can hold their own love and light to arrest this pattern. They remember and know they are loved and valued for the love and light they already are and how that love lights the world! Divine Human bio-light integration is where compassion and goodness replace any temporary conditional or time-split behaviors because humans had to do the harsh work of grow feelings into soul emotions so spirit could rebirth/re-code its heart essence passion and sensuality back into its multi-D nature.

The release of this pattern is also a signal that your old electrical system is transferring over to a neural arch of **GEM light web** which continually passes through the: thyroid carbon cell> to pineal crystal cell>to diamond cell pituitary>to plasma-particle cell hypothalamus> and back through heart spiral. The cellular rebirth of your multi-D light systems restores a natural built in boundary integrity of Self/ other, so soul sustains it unique free will expression within the context of the Oneness. This allows soul and spirit **diversity within unique soul-spirit multiplicity in the new DNA codes**. Your super-conscious sensitivities are now your greatest gift; for behind them lies true meta-essence, creative passion, and all the new divine quantum senses that have growing since you created yourself to BE. Hence, empathic distress or hyper-bio-cell sensitivity at the core essence level, is replaced with a light network of sensors. These provide your own bio-magnetic immunity and natural biosphere light protection via light vessel stabilization. Here light vessel's vacuum field of full spectrum light anchors you in **heart's gravity.**

**Multi-quantum sense communication** offers quantum free energy applications and solutions replace old human ways of linear space-time. For example, you've finally self-realized that **true abundance** is joy. True abundance is expression of your natural essence. True abundance is an exchange of creativity. True abundance is taking care of self and allowing self-love first. True creativity was meant to be the means of exchange, not money. Money has been the mediator between human creativity and slavery, just as religion has been the mediator between humans and their authentic Divine Self. Multi-quantum senses, and quantum Divine-heart communication access, replace old human mediation of linear space-time. Matter does not require mediation if matter is alive. **Quantum particles** twinkle off and on and pass-through solid matter. Why not see different beings in the painting moving about like in a movie each time your view it? Why not put a seed into a cup of water with a crystal and allow the squash to grow itself for you? Why not use your vessel as a biosphere/ mother ship, to visit the past or future races of other worlds? How about making a sculpture that allows its matter to sing? You easily accept such in your technology. It's time to feel your own creative matter come out of your own consciousness. As you allow matter to **change states**, many new potentials can be enjoyed. Plasma striates or new gravitation vectors of consciousness results in matter senses that feel like lava, gaseous, liquid or solid aggregates of new sensate or meta-material states. This allows for new molecular or particle/bio-essence experiences of light. This lays the groundwork for the gem vessels made of particle pure Essence light which will grow into your ascension vessels and inter-stellar biosphere vessels. Perhaps, you will see 3 or more of you in the market at the same time just for fun! So Beloved Masters a sweet Tribute to all your coming Divine Inspiration and Miracles.

It is quite wondrous to have the awesome privilege to share both *the core essence light body as the pure energy gem vessel or <u>Potential Body,</u> with your star families as a living art form.* You do this while still holding the holography of an entire theatrical universe re-birthing, re-imprinting/re-essence/ing, and-emerging as trillions of new star-sun multi-light verses. Already you are all feeling the cosmos calling home wanting to know how all our stories self-realized, as they came and went from love. Such aliveness is to feel the difference between creating out of limitation and creating out of a conscious heart, that naturally stream sources its own unlimited potentials. Being so delicately grateful, you're allowing yourself the stillness to remember and allow how **true creation works in the transitional light body into the ascended biosphere and the interstellar star-sun imprint vessel**. This authentic Presence-confidence allows you to enjoy and tans-sense this reality plane, till you grow into or walk out of this halo deck in, your glorious Rainbow Gem Vessel. Your new heart allows your consciousness to Breathe you and to sift through your potentials, as well as any debris from this plane, until the heart channels, dials in, or transmits what is an authentic potential for what you are ready to allow in to experience and expression. It can be quite a challenge at first, to watch your awareness to slide up and down in multi-streams of consciousness till a self-realized choice seemingly lands in the heart, and you hear its click or tone! *That moment is so clear, clean, and fulfilling that the matter is already manifested*. Be patient as you get the rust off creating and walk in the super feet of your Divine, now ordinary and extraordinary to fulfill, enjoy, and open up super conscious textures, senses, and qualities of essence that you haven't felt since you were in the angelic realm.

Contact inside self-love has landed in quantum-density. Bio-essence regeneration in new cosmic expressions has been born in the mangers of the New Earth universes. And, the inner fire of creative passion dissolves anything that is not love instantly. This is love's template in the matters of Life; and its matter lives again as new particles of joy, Creators! *** *2018-book exert from: The Free Energy Vessel (Traffordpress.com/bookstore)*